The Evolution of Technology

CAMBRIDGE HISTORY OF SCIENCE

Editors

GEORGE BASALLA
University of Delaware

WILLIAM COLEMAN
University of Wisconsin

Physical Science in the Middle Ages
EDWARD GRANT

Man and Nature in the Renaissance
ALLEN G. DEBUS

The Construction of Modern Science:
Mechanisms and Mechanics
RICHARD S. WESTFALL

Science and the Enlightenment
THOMAS L. HANKINS

Biology in the Nineteenth Century:
Problems of Form, Function, and Transformation
WILLIAM COLEMAN

Energy, Force, and Matter: The Conceptual Development of
Nineteenth-Century Physics
P. M. HARMAN

Life Science in the Twentieth Century
GARLAND E. ALLEN

The Evolution of Technology
GEORGE BASALLA

The Evolution of Technology

GEORGE BASALLA

Department of History, University of Delaware

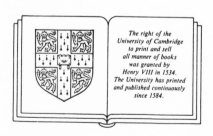
CAMBRIDGE UNIVERSITY PRESS

CAMBRIDGE

NEW YORK   PORT CHESTER   MELBOURNE   SYDNEY

For My Father

Published by the Press Syndicate of the University of Cambridge
The Pitt Building, Trumpington Street, Cambridge CB2 1RP
40 West 20th Street, New York, NY 10011, USA
10 Stamford Road, Oakleigh, Melbourne 3166, Australia

First published 1988
Reprinted 1989, 1990 (twice)

Printed in the United States of America

Library of Congress Cataloging-in-Publication Data

Basalla, George.
The evolution of technology.
(Cambridge history of science)
Bibliography: p.
Includes index.
1. Technology – History.   2. Technology – Philosophy.
1. Title.   II. Series.
T15.B29   1988   609   87-38217

British Library Cataloguing in Publication Data

Basalla, George
The evolution of technology.
1. Technological development, to 1987
I. Title
609

ISBN 0 521 22855 7 hardback
ISBN 0 521 29681 1 paperback

Contents

	Preface	*page* ix
I	Diversity, Necessity, and Evolution	1
II	Continuity and Discontinuity	26
III	Novelty (1): Psychological and Intellectual Factors	64
IV	Novelty (2): Socioeconomic and Cultural Factors	103
V	Selection (1): Economic and Military Factors	135
VI	Selection (2): Social and Cultural Factors	169
VII	Conclusion: Evolution and Progress	207
	Bibliography	219
	Sources of Quotations	241
	Index	245

Preface

This book presents a theory of technological evolution based on recent scholarship in the history of technology and relevant material drawn from economic history and anthropology. The organization and content of its chapters are determined by the nature of the evolutionary analogy and not by the need to provide a chronological account of events in the history of technology. However, because this study is primarily historical, not an exercise in the philosophy or sociology of technology, historical examples are used throughout to elucidate and support the theoretical framework. Major developments in the history of technology, such as the invention of the steam engine or the advent of the electrical lighting system, are introduced simultaneously with the unfolding of an evolutionary explanation of technological change.

The opening chapter announces three themes that reappear, with variations, in later sections of the work: *diversity* – an acknowledgment of the vast number of different kinds of artifacts, or made things, that have long been available; *necessity* – the belief that humans are driven to invent artifacts to meet basic biological needs; and *technological evolution* – an organic analogy that explains both the appearance and the selection of these novel artifacts. A close examination of these themes reveals that diversity is a fact of material culture, necessity is a popular but erroneous explanation of diversity, and technological evolution is a way of accounting for diversity without recourse to the idea of biological necessity.

The formal presentation of the theory of technological evolution begins in the second chapter, where it is established that an artifact is the fundamental unit for the study of technology, and that continuity prevails throughout the made world. The existence of continuity implies that novel artifacts can only arise from antecedent artifacts – that new kinds of made things are never pure creations of theory, ingenuity, or fancy.

If technology is to evolve, then novelty must appear in the midst of the continuous. Chapters III and IV survey the varied sources of novelty – the human imagination, socioeconomic and cultural forces, the diffusion of technology, the advancement of science – in primitive cultures as well as in modern industrialized nations. The conclusion drawn from this survey is that any society, at any time, commands more potential for technological innovation than it can ever hope to exploit.

Because only a small fraction of novel technological possibilities are sufficiently developed to become part of the material life of a people, selection must be made from among competing novel artifacts. Ultimately, the selection is made in accordance with the values and perceived needs of society and in harmony with its current understanding of "the good life." The selection process, and the forces that drive it, are covered in chapters V and VI.

In concluding, chapter VII addresses the issue of technological progress and human betterment. The traditional conception of progress is found to be internally flawed and incompatible with technological evolution. However, progress can be redefined so that it no longer conflicts with an evolutionary perspective.

Because a book of this breadth could not be written without the rich scholarly resources produced by historians of technology in the past several decades, I wish to acknowledge my debt to the authors listed in the bibliography. Specifically, I have made extensive use of the ideas and insights of George Kubler and Nathan Rosenberg.

I owe a special obligation to two close friends: William Coleman, my fellow editor for the Cambridge History of Science Series, who guided me in handling the evolutionary analogy; and Eugene S. Ferguson, my colleague at the University of Delaware, who counseled me on every aspect of the book. It is no exaggeration to say that this volume could not have taken its final form without their help.

Finally, I want to thank Catherine E. Hutchins of the *Winterthur Portfolio* for editing the text, Marie B. Perrone for typing the manuscript, Kenneth Marchionno for preparing the illustrations, and my wife and family for their long and consistent support.

CHAPTER I

Diversity, Necessity, and Evolution

Diversity

The rich and bewildering diversity of life forms inhabiting the earth has intrigued humankind for centuries. Why should living things appear as paramecia and hummingbirds, as sequoia trees and giraffes? For many centuries the answer to this question was provided by the creationists. They claimed that the diversity of life was a result and expression of God's bountiful nature: In the fullness of his power and love he chose to create the wonderful variety of living things we encounter on our planet.

By the middle of the nineteenth century, and especially after the publication of Charles Darwin's *Origin of Species* in 1859, the religious explanation of diversity was challenged by a scientific one. According to this new interpretation, both the diversification of life at any given moment and the emergence of novel living forms throughout time were the result of an evolutionary process. In support of Darwin's theories, biologists have proceeded to identify and name more than 1.5 million species of flora and fauna and have accounted for this diversity by means of reproductive variability and natural selection.

Another example of diversity of forms on this earth, however, has been often overlooked or too readily taken for granted – the diversity of things made by human hands. To this category belongs "the vast universe of objects used by humankind to cope with the physical world, to facilitate social intercourse, to delight our fancy, and to create symbols of meaning."[1]

Because distinct species cannot be identified with any precision among items of human manufacture, obtaining an accurate count of the different kinds of made things is difficult. A very rough

1

approximation of that figure can be reached by using the number of patents granted as an indicator of the diversity of the made world. In the United States alone more than 4.7 million patents have been issued since 1790. If each of these patents is counted as the equivalent of an organic species, then the technological can be said to have a diversity three times greater than the organic. Although faulty at several points, this attempt at measuring comparative diversification suggests that the diversity of the technological realm approaches that of the organic realm.

The variety of made things is every bit as astonishing as that of living things. Consider the range that extends from stone tools to microchips, from waterwheels to spacecraft, from thumbtacks to skyscrapers. In 1867 Karl Marx was surprised to learn, as well he might have been, that five hundred different kinds of hammers were produced in Birmingham, England, each one adapted to a specific function in industry or the crafts (Figure I.1). What forces led to the proliferation of so many variations of this ancient and common tool? Or more generally, why are there so many different kinds of things?

Our attempts to understand diversification in the made world, or even to appreciate its richness, have been hampered by the assumption that the things we make are merely so many instruments enabling us to cope with the natural environment and maintain the necessities of life. Traditional wisdom about the nature of technology has customarily stressed the importance of necessity and utility. Again and again we have been told that technologists through the ages provide humans with the utilitarian objects and structures necessary for survival.

Because necessity and utility alone cannot account for the variety and novelty of the artifacts fashioned by humankind, we must seek other explanations, especially ones that can incorporate the most general assumptions about the meaning and goals of life. This search can be facilitated by applying the theory of organic evolution to the technological world.

The history of technology, a discipline that focuses on the invention, production, and uses of material artifacts, benefits from the application of an evolutionary analogy as an explanatory device. A theory that explains the diversity of the organic realm can help us account for the variety of made things. This venture does have its pitfalls, however, as poet e. e. cummings warned, "A world of made is not a world of born."[2]

The evolutionary metaphor must be approached with caution because there *are* vast differences between the world of the made and the world of the born. One is the result of purposeful human activity, the other the outcome of a random natural process. One produces a sterile physical object, the other a living being capable of reproducing itself. Emphatically, I do not propose the establishment of a one-to-one correspondence between these markedly different domains. In the narrative and analysis that follow, I employ the evolutionary metaphor or analogy selectively, with the expectation that this metaphor will give us insights otherwise unavailable to the history of technology.

The nature of metaphor and its role in this book need additional clarification. Metaphors are not ornaments arbitrarily superimposed on discourse for poetic purposes. Metaphors or analogies are at the heart of all extended analytical and critical thought. Without metaphors literature would be barren, science and philosophy would scarcely exist, and history would be reduced to a chronicle of events.

Historians have long relied on metaphors in interpreting the past, especially organic metaphors invoking birth, growth, development, maturity, health, disease, senescence, and death. For the past century or so, those who specialize in the history of science and technology have routinely drawn upon a powerful political metaphor, that of revolution, to explain happenings in those areas. Thus, in suggesting that evolutionary theory be employed in understanding technological change, I am not introducing metaphor into a field that had never known the concept before; however, I am introducing a new metaphor and urging that its wider implications be considered seriously.

I ask that readers grant me the same indulgence they have extended to those who write about scientific and industrial revolutions. Just as historians of science and technology are not held responsible for all points of similarity between political revolt and radical scientific, technological, and industrial change, so I should not be taken to task if I do not draw parallels between every feature of the made and living worlds.

In one respect my use of metaphor differs from that of most historians: They utilize metaphors implicitly and often unconsciously; in this book I make explicit and conscious use of mine. Although our choice of, and approach to, metaphors may differ, we share the same aim – to make sense of the past.

Necessity

A well-known Aesop's fable is particularly relevant to the discussion of technology, diversity, and necessity. Once upon a time, wrote

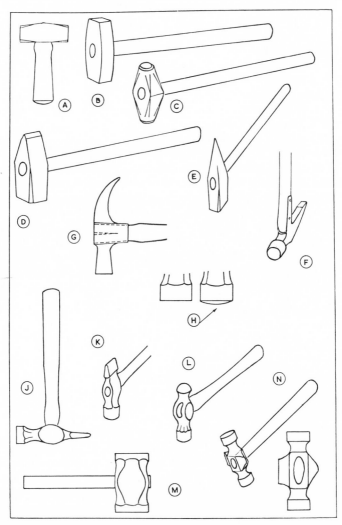

Figure I.1. Artifactual diversity as reflected in the forms of hammers used by English country craftsmen. I: A,B,C,D,E, – Stone mason's hammers used to break, cut, square, and dress stone; F,G – Carpenter's hammer with strengthened head; H – Curved hammer head, used to protect wood's surface when driving a nail; J – General woodworking hammer; K – Straight-peen blacksmith's hammer; L – Ball-peen, a general metalworking hammer; M – Chair-maker's hammer; N – Horse-

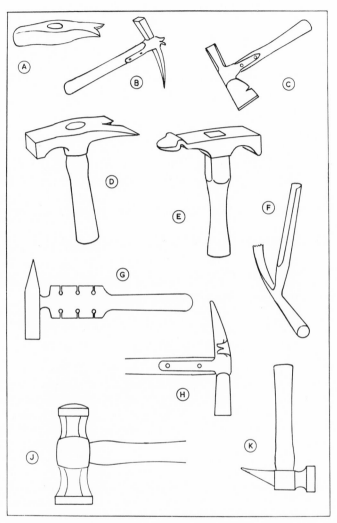

shoeing hammer (two views). II: A – Head of claw hammer used to withdraw nails; B – Slater's pick hammer; C – Lath hatchet; D – Cooper's nailing adze, used on barrel hoops; E – Butter firkin, used to open and close butter casks; F – Combination cheese-taster and hammer; G – Saw-sharpening and saw-setting hammer; H – Upholsterer's or saddler's hammer; J,K – Shoemaker's hammers. Source: Percy W. Blandford, *Country craft tools* (Newton Abbot, 1974), pp. 49, 55.

Aesop, a crow about to die of thirst came upon a tall pitcher partially filled with water. He tried again and again to drink from it, stooping and straining his neck, but his short beak could not reach the surface of the water. When he failed in an attempt to overturn the heavy vessel, the bird despaired of ever quenching his thirst. Then he had a bright idea. Seeing loose pebbles nearby, the crow began dropping them into the pitcher. As the stones displaced the water, its level rose. Soon the crow was able to drink his fill. The moral: necessity is the mother of invention. Modern commentators have elaborated on this message by praising those individuals who, when placed in seemingly impossible situations, do not despair but instead use wit and ingenuity to invent new devices and machines that solve the dilemma, meet basic biological needs, and contribute to material progress.

The belief that necessity spurs on inventive effort is one that has been constantly invoked to account for the greatest part of technological activity. Humans have a need for water, so they dig wells, dam rivers and streams, and develop hydraulic technology. They need shelter and defense, so they build houses, forts, cities, and military machines. They need food, so they domesticate plants and animals. They need to move through the environment with ease, so they invent ships, chariots, carts, carriages, bicycles, automobiles, airplanes, and spacecraft. In each of these instances humans, like the crow in Aesop's story, use technology to satisfy a pressing and immediate need.

If technology exists primarily to supply humanity with its most basic needs, then we must determine precisely what those needs are and how complex a technology is required to meet them. Any complexity that goes beyond the strict fulfillment of needs could be judged superfluous and must be explained on grounds other than necessity.

In surveying the needs and techniques essential to human beings a modern commentator might ask: Do we need automobiles? We are often told that automobiles are absolutely essential, yet the automobile is barely a century old. Men and women managed to live full and happy lives before Nikolaus A. Otto devised his four-stroke internal combustion engine in 1876.

A search for the origins of the gasoline-engine-powered motorcar reveals that it was not necessity that inspired its inventors to complete their task. The automobile was not developed in response to some grave international horse crisis or horse shortage. National leaders, influential thinkers, and editorial writers were not calling for the replacement of the horse, nor were ordinary citizens anxiously hoping that some inventors would soon fill a serious societal and personal need for motor transportation. In fact, during the first decade of existence, 1895–1905, the automobile was a toy, a plaything for those who could afford to buy one.

The motor truck was accepted even more slowly than the automobile. The success of military truck transportation during World War I combined with an intensive lobbying effort by truck manufacturers and the Army after the war finally resulted in the displacement of the horse-drawn wagon and, at a later date, the railroad. But the motor truck was not created to overcome obvious deficiencies of horse- and steam-powered hauling. As was the case with automobiles, the need for trucks arose after, not before, they were invented. In other words, the *invention* of vehicles powered by internal combustion engines gave birth to the *necessity* of motor transportation.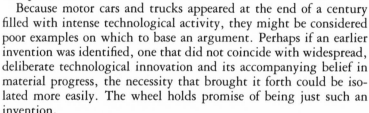

Because motor cars and trucks appeared at the end of a century filled with intense technological activity, they might be considered poor examples on which to base an argument. Perhaps if an earlier invention was identified, one that did not coincide with widespread, deliberate technological innovation and its accompanying belief in material progress, the necessity that brought it forth could be isolated more easily. The wheel holds promise of being just such an invention.

The Wheel

Popularly perceived as one of the oldest and most important inventions in the history of the human race, the wheel is invariably listed with fire as the greatest technical achievement of the Stone Age. In comic strips and cartoons, stone wheels and fire are portrayed as joint creations of prehistoric cave dwellers. This familiar portrayal, which first appeared in the late nineteenth century, is currently exemplified by the *B.C.* comic strip.

Those who have a better knowledge of the early history of human culture know that the origins of fire and the wheel do not date to the same time period. Fire has been in use for at least 1.5 million years, whereas the wheel is more than 5,000 years old. Even at this level of historical understanding, however, there is a tendency to pair the two items, placing them in a special category above and beyond all other human accomplishments. For example, when distinguished economic historian David S. Landes assessed the significance of the mechanical clock recently he conceded that it was "not in a class with fire and the wheel"[3] and hence deserved a lower ranking.

Whatever the degree of historical sophistication, most people believe that the use of wheeled transportation is a signal of civilization. The two are thought to be so closely linked that the progress made by cultures has been judged by measuring the extent to which they have exploited rotary motion for transportation. By that standard, to be without the wheel altogether is sufficient to set a culture apart from the civilized world.

In searching for the origins of this wonderful invention, there is no need to explore nature's realm. With the exception of a few microorganisms, no animal propels itself by means of a set of organic wheels spinning freely on axles. The source of the wheel must be sought among made things.

Before the coming of the wheel, large heavy objects were moved on sledges — wooden platforms with or without runners. Cylindrical rollers (smoothed logs) placed beneath the vehicle were used to facilitate the movement of the sledges, and it is thought that these rollers inspired the invention of the wheel.

Whatever the inspiration, wheels made their initial appearance in the fourth millennium B.C. across a broad area extending from the Tigris to the Rhine rivers. Current archaeological findings indicate that wheeled vehicles were invented in Mesopotamia and from there diffused to northwest Europe within a very short time. The first wheels were either solid wooden disks cut from a single plank or tripartite models consisting of three wooden slabs trimmed to shape and fastened together with cleats.

A strict reading of the archaeological record suggests that the first wheeled vehicles were used for ritualistic and ceremonial purposes. The earliest illustrations show them being used to carry effigies of deities or important persons. The oldest remains of wheeled conveyances are found in tombs; such vehicles, interred with the deceased as part of a religious burial ceremony, have been uncovered at various sites in the Near East and Europe.

Vehicles buried with the dead were often of the type used on the battlefield. Thus the ritualistic and ceremonial uses of the wheel were closely related to its employment in war. Military requirements exerted a powerful influence upon the subsequent development of wheeled vehicles. For example, pictorial and physical evidence supports the idea that the four-wheeled "battle wagon" and the two-wheeled "straddle car" (a chariotlike vehicle) of Mesopotamia were used early as moving platforms from which javelins could be hurled. The innovative spoked wheel, which demanded a high level of craftsmanship, was first utilized on war chariots in the second millennium

B.C. to create light and fast-moving vehicles that could be maneuvered easily during battle.

In addition to ritualistic and military uses, the wheel was also used in transporting goods. Although this third function is not directly recorded in the earliest archaeological evidence, we assume that wheeled vehicles could be, and were, used for more mundane purposes at an early date. Documentary evidence of wagons transporting farm goods such as hay, onions, reeds dates from 2375 to 2000 B.C., about a thousand years after the wheel's initial appearance. However, this time lag may simply reflect the ritualistic, ceremonial, and military nature of much of our arachaeological evidence. Despite the lack of strong proof for the transport function of wheeled vehicles in earliest times, it can be argued that the utilitarian aspect of the wheel was primary and that the necessity of transporting farm goods was the source of the invention of the wagon and cart.

Our discussion of the wheel and its uses has been confined to a relatively small geographical area. The story of the wheel in the rest of the world remains to be told. Wheeled vehicles appeared in India in the third and in Egypt and China in the second millennium B.C. As for Southeast Asia, Africa south of the Sahara, Australasia, Polynesia, and North and South America, people in those vast regions managed to survive, and in many cases prosper, without the help of the wheel. Not until modern times was rotary motion for transportation purposes introduced into these lands.

Especially interesting is the case of Mesoamerica (roughly Mexico and Central America). Although wheeled transport was unknown there prior to the arrival of the Spanish, Mesoamericans did make miniature wheeled objects. From the fourth to the fifteenth centuries A.D., clay figurines of various animals were fitted with axles and wheels to make them mobile (Figure I.2). Whether these figurines were toys or cult or votive objects is unknown; however, irrespective of their purpose, they show that the mechanical principle of the wheel was thoroughly understood and applied by people who never put it into use for transporting goods.

How are we to explain this failure to exploit an invention commonly held to be one of the two greatest technical achievements of all time? If we assume we are dealing with a people whose intellectual development was so stunted that they were unable to make practical use of the wheel, how can we account for the fact that they were capable of independently inventing the wheel in the first place? And how do we explain the flowering of the Aztec and Maya cultures with their many accomplishments in the arts and sciences?

The answer to these questions is simple. Mesoamericans did not

Figure I.2. Wheeled clay figurine made by the Aztecs (Mexico). Animal figurines employing the principle of the wheel and axle are found throughout Mesoamerica. They date from ca.A.D. 300 to the coming of the Spanish in the sixteenth century, a period when there was no wheeled transportation in the region. Source: Stuart Piggott, *The earliest wheeled transport* (Ithaca, N.Y., 1983), p. 15. Neg. no. 326744; courtesy Department of Library Services, American Museum of Natural History.

use wheeled vehicles because it was not feasible to do so given the topographical features of their land and the animal power available to them. Wheeled transport depends on adequate roads, a difficult requirement in a region noted for its dense jungles and rugged landscape. Large draft animals capable of pulling heavy wooden vehicles, were also needed, but Mesoamericans had no domesticated animals that could be put to that use. Men and women of Mexico and Central America traveled along trails and over rough terrain carrying loads on their backs. It was unnecessary to build roads for these human carriers of goods.

An even more persuasive case can be made against the universal superiority and applicability of the wheel by returning to its place of origin in the Near East. Between the third and seventh centuries A.D., the civilizations of the Near East and North Africa gave up wheeled vehicular transportation and adopted a more efficient and speedier way of moving goods and people: They replaced the wagon and cart with the camel. This deliberate rejection of the wheel in the very region of its invention lasted for more than one thousand years. It came to an end only when major European powers, advancing their imperialistic schemes for the Near East, reintroduced the wheel.

The camel as a pack animal was favored over wheeled transportation for reasons that become evident when the camel is compared with the typical ox-drawn vehicle. The camel can carry more, move faster, and travel farther, on less food and water, than an ox. Pack camels need neither roads nor bridges, they can traverse rough ground and ford rivers and streams, and their full strength is devoted to carrying a load and not wasted on dragging a wagon's deadweight. Once the camel and ox are compared, one wonders why the wheel was ever adopted in that region in the first place. A large share of the burden of goods in the Near East was always carried by pack animals. A bias for the wheel led Western scholars to underrate the utility of pack animals and overemphasize the contribution made by wheeled vehicles in the years before the camel replaced the wheel.

The more we learn about the wheel, the clearer it becomes that its history and influence have been distorted by the extraordinary attention paid to it in Europe and the United States. The Western judgment that the wheel is a universal need (as crucial to life as fire) is of recent origin. Fire, not the wheel, was the precious gift Prometheus stole from the gods and bestowed upon humanity. Similarly, fire, and not the wheel, was traditionally portrayed as the great civilizing agent in the literary and visual arts of Western culture. Not until the late nineteenth and early twentieth centuries did popular writers on technology elevate the wheel to the premier place it holds today.

This history of the wheel began as a search for a significant technological advancement that was produced in response to a universal human need. It has ended with the wheel seen as a culture-bound invention whose meaning and impact have been exaggerated in the West. Although this review is not meant to detract from the real importance of the wheel in modern technology, it does raise serious doubts about using it as a criterion to evaluate other cultures.

By putting wheeled transport into a broader cultural, historical, and geographical perspective, three important points emerge: First, wheeled vehicles were not necessarily invented to facilitate the movement of goods; second, Western civilization is a wheel-centered civilization that has carried rotary motion in transportation to a high state of development; and, third, the wheel is not a unique mechanical contrivance necessary, or useful, to all people at all times.

Fundamental Needs

The pursuit of need and invention has revealed that necessity is a relative term. A necessity for one people, generation, or social class may have no utilitarian value or may be a superficial luxury for another people, generation, or social class. At the same time that Europeans were energetically advancing wheeled transportation, Near Easterners were abandoning their experiment with the wheel, and Mesoamericans were adapting rotary motion to clay figurines. The story of the comparative reception and use of the wheel could be repeated for the other so-called necessities of modern life. Far from fulfilling universal needs, they derive their importance within a specific cultural context or value system.

This arouses the suspicion that it might be possible to strip away the false necessities, the trivial ones to which we have merely become accustomed, to reveal a core of fundamental needs applicable to humans living in any age and place. These universal needs would provide a firm ground on which to base an understanding of culture, including technology.

According to functionalist anthropologists and sociobiologists, every aspect of culture, material and nonmaterial, can be traced directly to the satisfaction of a basic need. In their view culture is nothing more than humanity's response to the fulfillment of its nutritive, reproductive, defensive, and hygienic needs. Critics of the biological theory, however, have proposed a number of strong counterarguments. Some have noted that phenomena central to culture, such as art, religion, and science, have very tenuous connections to human survival. Likewise, agriculture and architecture, which supposedly can be linked to the need for nutrition and shelter, manifest themselves in ways only remotely explicable by biological necessity. Modern agribusiness, for example, is motivated by much more than the concern for providing nourishment to humanity; a skyscraper is not simply a structure to protect people from the vagaries of the weather.

Some scholars argue that language is the most important feature of culture and that language, not biology, determines our definition of what we consider to be necessary or utilitarian. In their estimation, necessity is not something imposed by nature upon humanity but is a conceptual category created by cultural choice. Both sets of critics acknowledge external material constraints on culture; however, those constraints are seen as remote and of minor importance when compared with the immense range of cultural possibilities open to humankind. Biological necessity operates negatively and at extreme limits. It decrees what is impossible, not what is possible.

Another critical approach to theories of culture based on pre-existing fundamental needs evaluates the role of technology in the animal kingdom. Its proponents conclude that no technology whatsoever is required to meet animal needs. Proof of this assertion is found by observing the animal realm where the necessities of life are procured without the intervention of technology. Unlike the crow in Aesop's fable, birds in real life do not obtain water by resorting to elaborate technological stratagems. Birds and other animals do not dig wells or construct canals, aqueducts, and pipelines. Nature provides water, food, and shelter to them directly without any intervening made structures. Of course, some animals use sticks, stones, and leaves as crude tools for gathering food and as weapons for defending themselves, but animal tool behavior is so rudimentary and limited that it can scarcely be compared with the technology of the simplest of human cultures. There are no fire-using animals nor are there animals that routinely fashion new tools, improve upon old tool designs, use tools to make other tools, or pass on accumulated technical knowledge to offspring.

Given these facts, it is misleading to connect animal tool use to human technology by means of a smooth curve of transition. Even the earliest and crudest tools produced by humans imply considerable foresight and a level of mentality that sets them apart from the most sophisticated tools made by animals. As Karl Marx pointed out, the worst human architect is superior to the best insect nest or hive builder because only humans are able to envision structures in their imagination before erecting them.

Animals exist and thrive without fire or the simplest shaped stone utensils. Insofar as we are animals, on the zoological plane of existence, we too could live without them. Of course, without technology we could neither occupy nor visit many regions of the earth we now inhabit. Nor could we do most of the things we do in our everyday lives. But we could survive, and survival is what we have in mind when we ask how elementary a level of technology is required to meet our basic needs.

Because technology is not necessary in meeting the animal needs of humans, philosopher José Ortega y Gasset defines technology as the production of the superfluous. He remarks that technology was just as superfluous in the remote Stone Age as it is today. Like the rest of the animal kingdom we, too, could have lived without fire and tools. For reasons that are obscure, we began to cultivate technology and in the process created what has come to be known as human life, the good life, or well-being. The struggle for well-being certainly entails the idea of needs but those needs are constantly changing. At one time need prompted the building of pyramids and temples, at another time it inspired movement about the earth's surface in self-propelled vehicles, journeys to the moon, and the incineration and irradiation of entire cities.

We cultivate technology to meet our perceived needs, not a set of universal ones legislated by nature. According to French philosopher Gaston Bachelard the conquest of the superfluous gives us a greater spiritual stimulus than the conquest of the necessary because humans are creations of desire, not need.

A perceived need often coincides with an animal need, like the requirement for nourishment. Nevertheless, we must not lose sight of the fact that humans have now chosen an excessively complex, technological means of satisfying basic necessities. Instead of relying on nature directly for sustenance, we have devised the wholly unnecessary techniques of agriculture and cooking. They are unnecessary because plants and animals are able to grow and even thrive without human intervention, and because food need not be processed by fire before it is fit for human consumption. Agriculture and cooking are not prerequisites for human survival; they only become necessary when we choose to define our well-being as including them.

Humans have a different relationship with the natural world than do animals. Nature simply and directly sustains animal life. For humans, nature serves as a source of materials and forces that can be utilized in pursuit of what they choose to call for the moment their well-being.

Because the resources of nature are varied, and because human values and tastes differ from culture to culture, from time to time, and from person to person, we should not be surprised to find a tremendous diversity in the products of technology. The artifacts that constitute the made world are not a series of narrow solutions to problems generated in satisfying basic needs but are material manifestations of the various ways men and women throughout time have chosen to define and pursue existence. Seen in this light, the history of technology is a part of the much broader history of human aspirations, and the plethora of made things are a product of human minds replete with fantasies, longings, wants, and desires. The artifactual world would exhibit far less diversity if it operated primarily under the constraints imposed by fundamental needs.

Organic–Mechanical Analogies

Explaining artifactual diversity by means of a theory of technological evolution requires that we compare living organisms and mechanical devices. Such analogical thinking is a modern phenomenon with few precedents in antiquity. Aristotle, who wrote extensively on biological matters, made little use of mechanical analogies in his explication of the organic world. Not until the Renaissance did European thinkers begin to draw parallels between the organic and the mechanical. This association of what had hitherto been thought to be disparate elements was the result of the appearance of a host of new technological contrivances and the emergence of modern science.

Initially the flow of organic–mechanical analogies moved from technology to biology. Structures and processes in living organisms were described and explained in mechanical terms. In the middle of the nineteenth century there occurred a movement of metaphors in the opposite direction. The counterflow of metaphor was of critical importance; for the first time the development of technology was interpreted through organic analogies.

Widespread industrial growth, the geologist's ability to establish the antiquity of the earth, and the appearance of the Darwinian theory of evolution facilitated the application of organic analogies to the technological realm. This new mode of metaphorization had its most notable and lasting affects upon literature and anthropology. The literary uses of the organic–mechanical metaphor can be conveniently studied in the writings of Samuel Butler, the anthropological in the work of General Augustus Henry Pitt-Rivers (original surname Lane-Fox). Both of these men lived in mid-Victorian England and both were deeply influenced by Charles Darwin's *Origin of Species*.

In his utopian novel *Erewhon* (1872) and essays such as "Darwin Among the Machines" (1863) Samuel Butler whimsically explored the idea that machines developed in a fashion remarkably similar to the evolution of living beings. His ideas inspired the popular evolutionary fantasy novels of nineteenth- and twentieth-century science fiction in which rapidly evolving machines surpass and supplant humans whose own evolutionary development has stagnated. Butler's influence is also evident in modern speculative essays that predict either the coming of a new symbiotic relationship between humans and machines or the supersession of humankind by new forms of technology that are capable of self-replication, such as robots and computers.

Victorians proud of their industrial accomplishments were warned by Butler that it was to their advantage to pause and contemplate the wider implications of technological change. Machines, he said, have undergone a series of very rapid transformations from the simple stick wielded by our early ancestors to the steam engine of today. This development in the direction of greater complexity raises the possibility of the addition of a mechanical kingdom, comprised of all forms of mechanical life, to the existing plant and animal kingdoms.

Identifying machines as a new class of living beings would allow Victorians to arrange them into genera, species, and varieties, suggested Butler, and proceed from this classificatory exercise to the construction of an evolutionary tree illustrating the connections between the various forms of mechanical life. Darwin's theory, therefore, is perfectly compatible with the mechanical kingdom. The history of technology is filled with examples of machines slowly changing over time and replacing older models, of vestigial structures remaining as parts of mechanisms long after they had lost their original functions, and of machines engaged in a struggle for survival, albeit with the help of humans. The animal or plant breeder who practices artificial selection by choosing certain specimens for propagation is doing precisely what the machine builder and the industrialist do with mechanical life when they plan a new technological venture.

To skeptics who objected that machines cannot be said to live and evolve because they are incapable of reproducing themselves, Butler responded that in the mechanical kingdom reproduction is accomplished in a different fashion. The propagation of mechanical life depends on a group of fertile contrivances, called machine tools, that are able to produce a wide variety of sterile machines.

A more pressing issue than reproduction, cautioned Butler, is the nature of the future relationships of humanity and the machine. Because machines are more powerful, accurate, dependable, and versatile than humans, and because machines are changing rapidly before our eyes, humans cannot help but fall back to second place in a world dominated by technology. Of course, we could try to put a stop to mechanical evolution but that would mean the destruction of every machine and tool, every lever and screw, every piece of shaped material. Because we cannot halt mechanical progress, we must resign ourselves, advised Butler, to assuming the status of servants to our superiors.

Butler's evolutionary speculations, presented in a literary tour de force, enabled him to display his wit and ingenuity, his ambiguous response to advances in technology and science, and his criticisms of popular theological and philosophical propositions. Pitt-Rivers, a career military officer who later devoted his life to ethnology and archaeology, approached technological evolution in an entirely different manner. His acceptance of the evolutionism of Darwin and Herbert Spencer grew out of his military experience and a desire to catalog, classify, and display his personal collection of primitive weapons and tools.

Assigned to test new rifles for the British army and prepare an instruction manual for their use in 1852, Pitt-Rivers became interested in the history of firearms. Through his research, he became aware of the gradual and progressive modification in firearm design that had resulted in the creation of ever more powerful and accurate rifles. At about the same time he began assembling a prehistoric artifact collection and investigating relics being unearthed in the British Isles and northern Europe. His contact with these diverse artifacts prompted him to consider the best way to organize them for study and eventual display. Should they be arranged geographically according to their place of origin or was there some more fruitful scheme of classification?

Natural history offered one mode for a classificatory system — the Linnaean ordering of the vegetable and animal kingdoms into genera, species, and varieties. In this system, form was more significant than geography. Because Darwin had shown that taxonomic studies could be made to yield great and fundamental truths about the nature of living things, Pitt-Rivers resolved to ignore the geographical, temporal, and cultural dimensions of artifacts, follow the lead of natural history, and arrange his collection in a series of sequences composed of closely related forms.

Spencer's assertion that the entire history of life was marked by a development from the simple to the complex, the homogeneous to the heterogeneous, inspired Pitt-Rivers to make these the guiding principles in his arrangement of artifacts. He placed them in sequences that began with the very simplest tool, weapon, or utensil and progressed step by small step to the most complex one. This method was more than a convenient way to impose order on the varied products of material culture. Because every artifact was thought to have originated as an idea in the mind of its original maker the sequences bound together the material and intellectual aspects of life. The progressive, continuous series of related artifacts served as proof of the evolution of human culture from its primitive condition to the highest states of civilization.

Pitt-Rivers confined his collecting and classifying labors to preindustrial artifacts, and deliberately avoided the difficulty of dealing with the more complex and sophisticated products of Victorian technology. His focus on the primitive stemmed from a belief that the study of the simplest artifacts would reveal the thought processes of prehistoric men and women and clearly demonstrate the progressive nature of material culture. But, modern critics will interject, the primitive cannot be equated with the prehistoric: We have no right to suppose that the culture of modern Australian aborigines bears any resemblance to Paleolithic culture. Pitt-Rivers and other nineteenth-century evolutionary anthropologists would reply that at any time in history the many societies scattered across the earth reflect the different evolutionary stages through which all human culture has passed. They believed that each culture followed a single, broad course of evolutionary change with only minor deviations. If Australian aborigines used stone tools, then they were at precisely the same stage of cultural development that Paleolithic man had reached hundreds of thousands of years earlier.

Given these assumptions about cultural evolution and artifacts, Pitt-Rivers was not particularly interested in garnering rare or exotic specimens for his collection. Nor was he concerned with accurately dating his artifacts and placing them within a specific cultural context. Instead he searched for forms that filled in the gaps of existing sequences or that could be used to initiate new sequences (Figure I.3 and I.4). The overriding criterion in every case was how well a specimen fit in between two others in a sequence − that is, how much it contributed to the establishment of a continuous transition. In the organic as well as in the technological realms, gaps in a sequence represented missing links that could be filled eventually. If it appeared that there were more artifactual than organic missing links, it was because the collection and classification of plants and animals had been going on for centuries whereas the organization and analysis of made things had barely begun.

Pitt-Rivers was careful not to overstate the case for technological evolution or to draw far-fetched analogies between living organisms and material objects. For example, he felt it was permissible to justify his interest in weaponry and the origins of warfare by linking them to the Darwinian struggle for existence. But humans use weapons in their struggle; the weapons themselves do not fight for survival. Nor are weapons or other artifacts capable of reproduction. Anticipating these objections, Pitt-Rivers introduced the idea of unconscious selection. Through the ages, without premeditation or design, humans had selected the artifacts best suited for certain tasks, rejected those less suited, and gradually modified the surviving artifacts so that they performed their assigned functions better. As a result, artifactual change was directed along a progressive path even though artisans were totally unaware of the long-range implications of the slight improvements they had introduced. In meeting

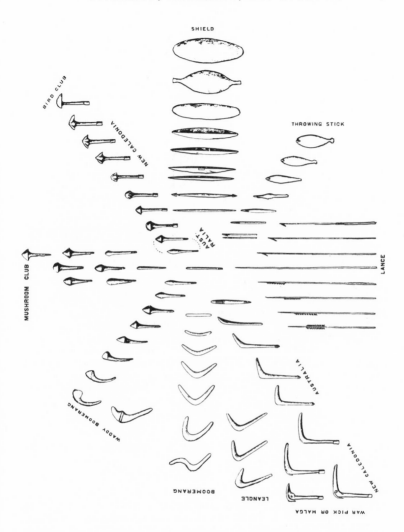

Figure I.3. The evolution of Australian aboriginal weapons. War clubs, boomer-angs, lances, throwing sticks, and shields were arranged by Pitt-Rivers so that they would appear as evolutionary sequences radiating out from the simple stick at the center. These are not historical sequences; all of the weapons displayed were in use in modern times. Pitt-Rivers assumed that the simpler artifacts, those located closer to the center, were "survivals" of earlier forms. Source: A. Lane-Fox Pitt-Rivers, *The evolution of culture* (Oxford, 1906), pl. III; reprinted by AMS Press, Inc., New York.

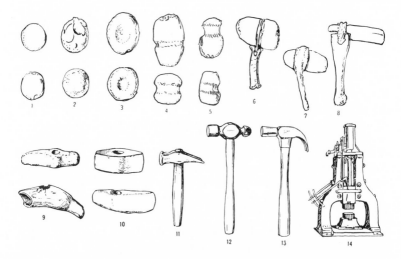

Figure I.4. The evolutionary history of the hammer, from the first crudely shaped pounding stone (1) to James Nasmyth's gigantic steam hammer of 1842 (14). This evolutionary sequence of a familiar hand tool was prepared by the staff of the U.S. National Museum to "indicate how the mind of man has arrived at certain datum points which mark epochs in progress." Following the example set by Pitt-Rivers (Fig. I.3), the "specimens are arranged in the order of their grade of development irrespective of race, place, or time." Source: Walter Hough, "Synoptic Series of Objects in the United States National Museum Illustrating the History of Inventions," *Proceedings of the United States National Museum* 60 (Washington, D.C., 1922), art. 9, p. 2, pl. 16.

an immediate need, they had inadvertently helped to promote technical progress.

For a modern observer to dismiss Pitt-Rivers's ideas as an over-enthusiastic and uncritical application of Darwinism to material culture is all too easy. We must remember that he witnessed at first hand the immense success of Darwin's theory in biology and that he was personally acquainted with some of the master's friends. That he might want to make his contribution to the spread of evolutionary doctrine is understandable. On the other hand, twentieth-century anthropologists and historians have rejected the belief that unilinear technological progress is a hallmark of human culture, and they have demonstrated the fallacy of supposing that the cultures of prehistoric and living primitive peoples are virtually identical. These modern criticisms, which seriously challenge key elements of Pitt-

Rivers's theories, have been widely publicized. Less well known are the original and enduring aspects of Pitt-Rivers's approach.

At a time when material culture studies were largely descriptive, if not outright antiquarian, Pitt-Rivers offered a theoretical basis for the integration of intellectual and technological achievements. An artifact was more than an inert object hastily fashioned to meet a need. It was a surviving remnant of the human mind that conceived it. Contrary to many of his contemporaries, Pitt-Rivers believed that technological change was not accomplished by a series of great, unrelated leaps forward by a few heroic inventors. Instead, the form of a modified artifact was based on that of a preexisting predecessor. From this followed the insight that every made thing could be placed within a sequence, which itself was interconnected to other sequences, and that, if we followed these backward in time, they would converge, leading us to the traces of the earliest human artifacts.

Cumulative Change

Butler and Pitt-Rivers were by no means representative of the prevailing view of the nature of technological change. The evolutionary, or continuous, explanation that they adopted was much less widely accepted than the revolutionary, or discontinuous, interpretation. According to the latter, inventions emerge in a fully developed state from the minds of gifted inventors. In this heroic theory of invention, small improvements in technology are ignored or discounted and all emphasis is placed upon the identification of major breakthroughs by specific individuals – for example, the steam engine by James Watt, or the cotton gin by Eli Whitney.

Not long after Darwin published *Origin of Species*, Karl Marx, a great admirer of the English naturalist, called for a critical history of technology written along evolutionary lines. He believed such a history would reveal how little the Industrial Revolution owed to the work of individual inventors. Invention is a social process, argued Marx, that rests on the accumulation of many minor improvements, not the heroic efforts of a few geniuses.

In the first half of the twentieth century, the heroic view of invention was challenged by three American scholars – William F. Ogburn, S. C. Gilfillan, and Abbott Payson Usher – who advanced theories of technological change that drew upon Darwinism. Ogburn, a sociologist and the most influential of the three, began by defining invention as combining existing and known elements of culture in order to form a new element. The outcome of this process is a series of small changes, most of them patentable, but none of them constituting a sharp break with past material culture.

Ogburn claimed that a fixed percentage of individuals with superior inventive ability can be found among all peoples. As the population grows in any country, the number of potential inventors increases proportionally. If these inventors happen to be born into a culture that provides technical training and places a premium on novelty, then inventions are bound to appear in quantity. Initially, the pace of innovation is slow as a stockpile of inventions is established. The subsequent accumulation of novelties stimulates innovation because the number of elements available for combination has grown. Soon the accumulated novelties reach a critical point and a chain reaction takes place greatly accelerating the rate of inventive activity.

Ogburn made no attempt to test his highly abstract theory by determining if it was in agreement with a sizable body of empirical evidence. In contrast, his fellow sociologist S. C. Gilfillan, in the 1930s, wrote companion volumes on invention – the first offering a sociology of invention, the second a detailed study that focused on the evolution of the ship from its origin as a floating log to the modern diesel-engine motorship.

Gilfillan was adamantly opposed to any theory of technological change that assigned inventions to what he called "titular inventors," those whose names were enshrined in the popular mythology of invention. Adhering to the Darwinian model, he wrote of the "undivided continuum of inventional reality"[4] and blamed language, custom, and social conventions for breaking down the continuum into a series of discrete, identifiable inventions.

The test of Gilfillan's theory is found in his second volume. According to him, the ship began as a hollowed-out log that was paddled by hand. When the earliest sailors stood up in their dugout canoes and found that the wind blowing against their garments increased the speed of their vessels, the sail was invented. Reconstructing the entire history of sailing ships from that point, using an evolutionary perspective, is relatively easy. Only the steam-powered craft seemingly disrupts the continuous flow. Gilfillan overcame this obstacle by placing the origin of the steamship in the Byzantine Empire. A war vessel moved by paddle wheels powered by three pair of oxen appears in an illustration from the early sixth century A.D. Paddle-wheel boats using ox or horse power thereafter evolved in a regular fashion. In the eighteenth century Europeans and Americans substituted steam for animal power to turn the paddle wheels. The issue was not the steam engine versus the sail, but the use of a steam engine versus oxen and horses to power a paddle-wheel boat.

Gilfillan concedes that there may be a dozen or so maritime inventions that could be termed abrupt in that they had no known or obvious predecessor. The ancient oxen-driven paddle-wheel boat is one such anomaly. Given that the development of the ship necessitated the accumulation of hundreds of thousands of minor inventions, Gilfillan is not troubled by the few innovations that appear to contradict his evolutionary stand. He maintains that the anomalies can be explained if we recognize that the cumulative process did not always take place in public, with the building of full-scale vessels. Gradual improvements may have been made in a series of rough sketches, formal drawings, or models before the results were tried in a full-size working ship. In this fashion abrupt inventions in the evolution of the ship can be dispensed with and Gilfillan's continuous curve of change reinstated.

Economic historian Abbott P. Usher, however, found the theories of invention put forth by Ogburn and Gilfillan excessively mechanistic. Inventors were depicted as mere instruments in a rigidly predetermined historical process. By emphasizing the social character of invention, and calling attention to the cumulative effects of small improvements, the two had ignored the importance of the individual inventor's efforts and insights. They would have us believe, Usher argued, that when the critical number of novel elements is reached the invention will appear automatically, with only a little help from an inventor.

Therefore, Usher proposed the *cumulative synthesis approach* to invention, an approach that modified the continuous explanation and enriched it with the findings of Gestalt psychology. Usher's theory contained four premises.

1. *Perception of the problem* – an incomplete or unsatisfactory pattern in need of resolution is recognized.
2. *Setting the stage* – data related to the problem is assembled.
3. *Act of insight* – a solution to the problem is found by a mental act that is not predetermined. This act goes beyond the *act of skill* normally expected of a trained professional.
4. *Critical revision* – the solution is fully explored and revised (with possible refinements made because of new acts of insight).

Central to Usher's thesis are the acts of insight that essentially solve the problem. These acts are as important to major, or strategic, inventions as they are to minor ones. The cumulative synthesis of lesser individual inventions eventually produces the strategic inventions better known to history. Yet the process is neither automatic nor predetermined. Sheer numbers of inventions do not guarantee that a major technological change will occur. The key is always the inventor's act of insight by which certain elements are chosen, combined in innovative ways, and made to yield a solution.

The acts of insight might be probed by psychologists, but they are, for the most part, inexplicable. They introduce the role of the mental faculties to the process of invention and by their presence indicate precisely at what point economic forces can be brought to bear. When the stage is being set (step 2) and the solution critically revised (step 4), economic intervention is likely to be effective. The acts of insight (step 3), on the other hand, are unresponsive to economic influence. They belong to the psychological, not the economic, realm.

Even though Usher came to study the inventive process as an economic historian, his theory transcended a strict economic or social explanation. By stressing the psychological aspects of invention, he served notice that the emergence of novelty must be dealt with in a broader context. Economists and economic historians who currently study invention do not follow Usher's lead on the importance of the acts of insight. Many of them do, however, accept his Darwinian-inspired idea that technical progress is the result of cumulative change.

A Modern Theory of Technological Evolution

My survey of past attempts to explain technological change by use of an evolutionary model has laid the groundwork for a consideration of the theory I will develop in this book. The study of Butler and Pitt-Rivers revealed that artifacts, like plant- and animal-life forms, can be arranged in continuous, chronological sequences. However, a modern theory of technological evolution cannot be built on an evocation of Darwinism for the purposes of literary and social satire (Butler), or on hypothetical chains of related primitive weapons (Pitt-Rivers). It is likewise unsatisfactory to limit the choice of illustrative examples to a single field of technology (Gilfillan), or to pursue a highly theoretical approach and ignore the technical details of artifactual change (Ogburn). Therefore, my theory will be supported throughout by detailed case studies of artifacts chosen from diverse technologies, cultures, and historical eras.

Butler, Pitt-Rivers, Gilfillan, Ogburn, and Usher all stressed the accumulation over time of small variations that finally yielded novel artifacts. Usher, by introducing "acts of insight" into the inventive process, drew attention to the role of individual creativity but he remained convinced that major inventions resulted from the cumulative synthesis of a series of minor ones. In the cumulative theory of invention change is slow and inevitable, and there is little room for the bold innovations of gifted individuals. My theory of technological evolution recognizes the larger changes, often associated with name inventors, as well as smaller changes made over a long duration. Hence, I accept periods of rapid technological change and times of relative stability.

Anyone advocating the continuous nature of technological change must acknowledge, and account for, the popularity of the opposing discontinuous view. There are many people who believe that technology advances by leaps from one great invention to another as the genius inventor creates a host of wonderful inventions through sheer mental effort. I reveal the sources of this belief by examining the relevant ideas and institutions of Western civilization that fostered its origin and growth.

Finally, my theory of technological evolution, unlike any of its predecessors, is rooted in four broad concepts: diversity, continuity, novelty, and selection. As I have already shown, the made world contains a far greater variety of things than are required to meet fundamental human needs. This *diversity* can be explained as the result of technological evolution because artifactual *continuity* exists; *novelty* is an integral part of the made world; and a *selection* process operates to choose novel artifacts for replication and addition to the stock of made things. The remainder of this book will be devoted to a thorough analysis of the theoretical and artifactual ramifications of these four concepts.

CHAPTER II

Continuity and Discontinuity

Introduction

A large segment of the modern public believes that technological change is discontinuous and depends on the heroic labors of individual geniuses, such as Eli Whitney, Thomas A. Edison, Henry Ford, and Wilbur and Orville Wright, who single-handedly invent the unique machines and devices that constitute modern technology. According to this view inventions are the products of superior persons who owe little or nothing to the past.

The smaller scholarly community that concerns itself with issues in the history of technology and science rejects this explanation as simplistic because it reduces complex technological developments to a series of great inventions that precipitately burst upon the scene. However, some historians have offered more sophisticated formulations of the discontinuous explanation that do not rely upon the contributions of heroic inventors. Such theorists take their cue from the supposed revolutionary nature of scientific change.

Science, Technology, and Revolution

Recent scholarship in the history and philosophy of science has tended to favor the discrete character of scientific change. This outlook derives ultimately from the study of the emergence of modern science in the sixteenth and seventeenth centuries. Since the French Revolution, the work of Copernicus, Galileo, Kepler, and Newton has been described by the term *revolution*, a political metaphor that implies a violent break with the past and the establishment of a new order.

The political metaphor was applied not only to the arrival of a new way of studying nature but also to any substantial change within a science. Hence, reference is made to the astronomical, chemical and biological revolutions of the past; to revolutions initiated byHarvey, Bacon, Darwin, Mendel, or Einstein; or to twentieth-century revolutions in quantum physics, astrophysics, and molecular biology.

Scientific revolutions take on a special importance for the study of technological change when technology is placed in a subordinate position to science. This situation usually occurs when technology is erroneously defined as the application of scientific theory to the solution of practical problems. For, if technology is nothing more than another name for applied science, and if science changes by revolutionary means, then technological change too must be discontinuous.

Of course, science and technology have interacted at many points, and key modern artifacts could not have been produced without the theoretical understanding of natural materials and forces provided by science. Nevertheless, technology is not the servant of science.

Technology is as old as humankind. It existed long before scientists began gathering the knowledge that could be used in shaping and controlling nature. Stone-tool manufacture, one of the earliest known technologies, flourished for over two million years before the advent of mineralogy or geology. The makers of stone knives and axes were successful because experience had taught them that certain materials and techniques yielded acceptable results, whereas others did not. When a transition was made from stone to metal (the earliest evidence for metal working has been dated ca. 6000 B.C.), the early metal workers, in a similar fashion, followed empirically derived recipes that gave them the copper or bronze they sought. Not until the late eighteenth century was it possible to explain simple metallurgical processes in chemical terms, and even now there remain procedures in modern metal production whose exact chemical basis is unknown.

In addition to being older than science, technology, unaided by science, is capable of creating elaborate structures and devices. How else can we account for the monumental architecture of antiquity or the cathedrals and mechanical technology (windmills, water-wheels, clocks) of the Middle Ages? How else can we explain the many brilliant achievements of ancient Chinese technology?

The arrival of modern science did not put an end to endeavors that were primarily technological; people continued to produce technological triumphs that did not draw upon theoretical knowledge. Many of the machines invented during Great Britain's Industrial

Revolution had little to do with the science of the day. The textile industry, at the heart of eighteenth-century economic growth, was not the result of the application of scientific theory. The inventions of John Kay, Richard Arkwright, James Hargreaves, and Samuel Crompton, which were crucial to increased textile production, owed more to past craft practices than they did to science.

Only during the latter half of the nineteenth century did science begin to have a substantial influence on industry. Developments in organic chemistry made possible the establishment of large-scale synthetic dye production, and the study of the nature of electricity and magnetism laid the foundations for the electrical light, power, and transport industries. The twentieth century witnessed the further expansion of science-based technologies. Despite the influx of new scientific theories and data, modern technology involves far more than the routine application of the discoveries made by scientists. In modern industry science and technology are equal partners, each making its unique contributions to the success of the enterprise in which they are engaged. Even today, however, it is by no means exceptional for an engineer to devise a technological solution that defies current scientific understanding or for engineering activity to open up new avenues for scientific research.

The issue of discontinuous change within technology was revived by Edward W. Constant in *The Origins of the Turbojet Revolution* (1980). Drawing upon Thomas S. Kuhn's *The Structure of Scientific Revolutions* (1962), Constant argues that a discontinuity existed between turbojet engines, the major aircraft power source since World War II, and the older propeller–piston engines (Figure II.1). The turbojet, which has neither piston, nor cylinder, nor propeller, was not the evolutionary outcome of the continuous improvement of the propeller–piston engine.

Constant believes that the turbojet revolution deserves its name because the turbojet is a holistic system that has technological antecedents but differs from them radically; the design of the turbojet, and its incorporation into aircraft, called for the application of advanced scientific theories in aerodynamics; and the turbojet was created by a small group of men who were not part of the conventional aeroengine community. Of all of these factors, Constant stresses the importance of the emergence of a new community of technological practitioners associated with the turbojet. This new community was identical with neither the conventional aeroengine community nor the older community of steam- and water-turbine technologists. In Constant's terms, technological revolution becomes "the professional commitment of either a newly emerging community or redefined community to a new technological tradition."[1]

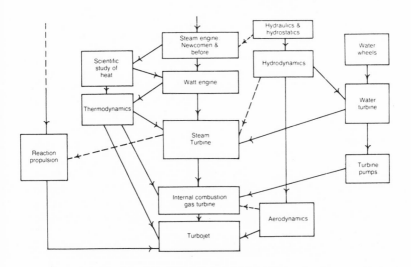

Figure II.1. E. W. Constant's diagram illustrating the relationship between artifact and theory in the lineage of the turbojet engine. The main artifactual stream, depicted in the center, moves from steam engines, to steam and internal combustion gas turbines, to the turbojet. On the far right water turbines and turbine pumps are linked to this main stream, and in the columns immediately adjoining the center the influence of relevant physical theories is indicated. Remember that these theoretical contributions manifest themselves in the form of some novel tangible thing whose antecedents predate the theory. Despite the existence of this artifactual network, Constant remained convinced that the turbojet was a revolutionary advance and not an evolutionary one. Source: Edward W. Constant II, *The origins of the turbojet revolution* (Baltimore, 1980), p. 4.

Two crucial assumptions underlie Constant's explanation: that technology is primarily knowledge; and that the community of technological practitioners is the fundamental unit to be studied in technological change. These assumptions deserve exploration.

Despite its seemingly revolutionary character the turbojet engine was not a machine without antecedents. The turbojet belongs to the two-hundred-year-old tradition of turbine development that encompasses water turbines, turbine water pumps, steam turbines, internal combustion gas turbines, piston engine superchargers, and turbosuperchargers. None of these has pistons and cylinders but they all have a turbine wheel with fins or buckets that, when acted upon by water, steam, or hot gases cause the wheel to rotate rapidly. Therefore, at the level of the artifact, two centuries of continuity has prevailed in the family of turbines, whatever their varied uses or energy sources.

The artifact – not scientific knowledge, nor the technical community, nor social and economic factors – is central to technology and technological change. Although science and technology both involve cognitive processes, their end results are not the same. The final product of innovative scientific activity is most likely a written statement, the scientific paper, announcing an experimental finding or a new theoretical position. By contrast, the final product of innovative technological activity is typically an addition to the made world: a stone hammer, a clock, an electric motor.

Historian Brooke Hindle has claimed that the artifact in technology occupies a position superior to that held by artifacts in science, religion, politics, or any of the intellectual or social pursuits. At every point technology is intimately involved with the physical, with the material; artifacts are both the means and the ends of technology. The three-dimensional physical object is as much an expression of technology as a painting or a piece of sculpture is an expression of the visual arts. The artifact is a product of the human intellect and imagination and, as with any work of art, can never be adequately replaced by a verbal description.

The centrality of the artifact for an understanding of technology is a key element of the evolutionary theory being developed here. In this chapter, as well as the rest of the book, the artifact is the primary unit for study. Artifacts are as important to technological evolution as plants and animals are to organic evolution.

Case Studies in Continuity

Given the primacy of the artifact in the study of technology, the continuous nature of technological change can best be established by using case studies to show precisely how key artifacts – such as the steam engine, the cotton gin, or the transistor – emerged in an evolutionary fashion from their antecedents. The following examples of artifactual change illustrate the evolutionary hypothesis despite the fact that initially they appear to be excellent candidates for use in supporting the contrary discontinuous explanation.

Stone Tools

The oldest surviving made things are stone tools. They stand at the beginning of the interconnected, branching, continuous series of artifacts shaped by deliberate human effort. Individual branches of the series may have stopped at cul-de-sacs but the wider stream of made things has never been broken. The modern technological world in all of its complexity is merely the latest manifestation of a continuum that extends back to the dawn of humankind and to the first shaped artifacts. Stone implements may not offer a crucial test for the evolutionary thesis, but they provide the best illustration of continuity operating over an extended period of time.

For at least two million years men and women in all parts of the earth made stone tools, hundreds of billions of them. These tools constitute the most ancient, widespread, and numerous artifacts in existence today. For most of the time tools were made by chipping and flaking techniques that when carried out by skilled workers, yielded a serviceable instrument within a relatively short time – a matter of minutes or hours. Axes, adzes, hammers, knives, and scrapers of all sorts were produced by these means. The Neolithic period, which began some eight thousand years ago, witnessed the introduction of agriculture, domestication of animals, and pottery, as well as polished stone tools made by the laborious process of pecking and grinding the stone to the desired shape and finish. Grinding, especially, required days or weeks of work; however, it resulted in tools well suited for prolonged hammering and chopping.

Whatever the technique employed, the form of stone tools changed very slowly over the long period of their use. Since the mid-nineteenth century, a number of archaeologists have patiently and ingeniously identified and dated implements that, to untrained eyes, seemed to be similar in shape, size, and material. Once this archaeological evidence was reassembled in chronological order, its most striking feature was the perfect continuity that was maintained for hundreds of thousands of years as wave after wave of different human cultures engaged in stone-tool making.

In the study of stone tools we search in vain for discontinuous jumps to wholly new forms. The long-lasting shapes of these artifacts even persisted as people made the radical shift from stone to copper and bronze. Traditionally, stone has been associated with primitivism, metal with civilization. As a material for making tools, stone does have weaknesses. Although it is easily obtained and worked, stone is less durable than metal and less readily shaped. The form of a stone tool is more closely determined by the nature of its material than is the form of a metal tool. The latter can be cast into almost any shape called for by the task at hand. The metal tool is less brittle, therefore, less likely to break. If it does break or show wear, it can be melted down and recast.

Someone unfamiliar with the subsequent history of tools might conclude from this comparison that the appearance of metal ushered in a new era in tool making. But, on the contrary, continuity prevailed. The earliest metal tools had as their closest antecedents stone prototypes. Eventually new metal tools did emerge, but the weight of the tradition of stone technology exerted a long-lasting influence on the shape of these tools. That influence is evident in the form of familiar modern tools, such as the axe, hammer, and saw, and is also apparent in electric and pneumatic tools, which preserve the principles and movements of early stone implements.

This example of continuity is dramatic yet it is open to criticism. Someone could argue that these stone tools owe their remarkable stability to the fact that at an early stage in their evolution they acquired the best possible form for the function they were to perform and hence did not change. Even if this were not true, one could make a case for excluding stone tools from this list of illustrative case studies on the basis that they represent an anomaly because they are so ancient and simple. For critics who would take exception to stone tools, I will present more persuasive tests of continuity in technology that utilize relatively complex machines, created by famous inventors in modern times, rather than artifacts fashioned anonymously in the prehistoric era.

The Cotton Gin

An investigation into the continuous development of a complex machine can profitably begin with Eli Whitney's cotton gin. Although many authors have written about Whitney and his revolutionary invention, far fewer have ventured to place his cotton gin within a continuous stream of artifacts.

According to the popular histories of invention, Whitney, a young New Englander with great mechanical talents, first encountered cotton and its processing problems while on a trip to a Georgia plantation in 1793. The source of the problems, short staple or Inland cotton, could be grown in most of the South but its fibers clung tightly to its green seed making it difficult to clean. A slave spent at least three hours cleaning one pound of cotton by tedious hand labor. Shortly after his arrival in Georgia, Whitney began working on a means to speed up the process. Having observed slaves manually separating fiber from seed, he visualized a machine that could duplicate the motions of their hands. Within a few days he had built a model of the gin that was to change cotton growing in the South forever.

Only the more scholarly studies of Whitney mention the fact that mechanical cotton gins were already in wide use in the South at the time of Whitney's visit. Such engines were capable of cleaning the long staple, or Sea Island, cotton, a plant with a limited growing range but with a fiber easily removed from its black seed. Because of the existence of these gins, it was unnecessary for the inventor to bridge the great gulf between the organic – fingers tugging to free stubborn fibers from seed – and the mechanical. That step had been accomplished much earlier in India where the first cotton cloth was produced centuries before the Christian era.

The Indian gin or *charka*, based on the roller principle, was itself a variant of a still older sugar cane press. The *charka* consists of a pair of long wooden cylinders set in a frame, pressed together, and rotated about the longitudinal axis by a crank. The rotating cylinders, which are fluted with a series of fine, lengthwise grooves, separate seed from fiber by squeezing the cotton boll as it passes between them.

This primitive gin was used wherever long staple cotton was grown and processed. By the early twelfth century the machine was known to Italian artisans as the *manganello*; it appeared in a Chinese illustration of the fourteenth century; and in the eighteenth century it was depicted in Diderot's *Encyclopédie*. In 1725 the roller gin was introduced into the Louisiana region from the Levant and by 1793 it was established in the cotton-growing South where Eli Whitney met it.

Whitney's challenge was to make a gin that would clean short staple cotton. His invention consisted of a rotating wooden cylinder into which were set regularly spaced rows of bent wire teeth whose shape is similar to the wire teeth used in various wool-carding devices. The teeth in Whitney's gin passed through a slotted metal breastwork with openings just wide enough to admit them and cotton fiber but not the cotton seed. Thus the seeds were trapped beneath the breastwork while the fibers were pulled upward and free. The cleaned cotton was then brushed from the teeth by the second rotating cylinder, which had rows of bristles. Like the centuries-old roller gin, Whitney's gin relied on a set of rotating cylinders. Unlike the older gin, Whitney's had a slotted plate (breastwork) to immobilize the seeds while they were being stripped of fiber.

This excursion into ancient cotton processing technology is not meant to prove that inventions are inevitable, or that the modern cotton gin was first constructed by some Indian artisans, or that Eli Whitney was less clever than we have been led to suppose. Whitney's gin cleaned short staple cotton, something the ancient roller gins could not do. Acknowledging the *charka*, however, shows that Eli Whitney's invention had artifactual antecedents whose overall structure and mechanical elements were adapted by the American inventor to suit his purposes (Figure II.2).

Not everyone who looked at the old roller gins visualized how to transform them into a machine capable of handling short staple cotton; some individuals had attempted to adapt the *charka* to processing Inland cotton before Whitney traveled South, but none met with success. Whitney's invention not only succeeded where others had failed but also served as the point of origin for an entirely new set of artifacts – a series of modern cotton gins. This new evolutionary series began almost immediately after Whitney's model was put to work. The widespread use of the Whitney-inspired gins owed as much to his inventive genius as it did to environmental, social, economic, and political conditions that favored the cultivation of cotton in America and elsewhere.

Several lessons can be drawn from the Whitney story. The obvious one is that Whitney's invention of the cotton gin was part of the evolutionary development of technology. Less obvious is the realization that all variants of an artifact are not of equal importance. Some are simply inoperable; some are ineffective; and some are effective but have little technological and social influence. Only a few variants have the potential to start a new branching series that will greatly enrich the stream of made things, have an impact on human life, and become known as "great inventions" or "turning points in the history of technology."

Recognition of the significance of Whitney's cotton gin depended on the growing demand at home and abroad for cheap cotton and the limited availability of slave and paid laborers to process the raw material manually. In a society dominated by woolen or linen cloth, or in one in which cheap manual labor was freely available, Whitney's machine would not have served as the prototype for a spate of more powerful and effective gins. In either of those alternative societies, the cotton gin would have been a mechanical curiosity without social, economic, or technological influence.

Thus, the significance of an invention cannot be determined solely by its technological parameters – it cannot be evaluated as if it were a thing unto itself. An invention is classified as "great" only if a culture chooses to place a high value upon it. Likewise, the reputation of its inventor is tied to cultural values. In either of the alternative worlds just described, Whitney would not be honored as a heroic inventor; he would be ignored or at best be looked upon as the eccentric builder of a trivial device.

Steam and Internal Combustion Engines

The cotton gin was the most important technological contribution to the growth of the American South's economy between 1790 and 1860. At about the same time, the steam engine played a somewhat comparable role in the British economy. Like the cotton gin, it, too, has been popularly viewed as a contrivance with virtually no history. In 1842 W. Cooke-Taylor, a commentator on the British industrial scene remarked: "The steam engine had no precedent . . . [it] sprang into sudden existence, like Minerva from the brain of Jupiter."[2] Or did it spring from the brain of James Watt?

Popular accounts tell us that young James Watt was inspired to invent the steam engine as he watched steam rising from the spout of a tea kettle (Figure II.3). The fanciful legend is undermined by the fact that working Newcomen steam engines existed in England at the very moment Watt was contemplating vapors from the boiling tea water. Some sixty years separate the appearance of Thomas Newcomen's working atmospheric steam engine in 1712 and the completion of a successful full-sized steam engine by Watt (1775). To complicate the matter further, Watt's version of the steam engine grew out of his dissatisfaction with a small-scale model of a Newcomen engine he was asked to repair.

The Newcomen engine utilizes the condensation of steam to create a partial vacuum beneath a piston, which is then forced down by the greater atmospheric pressure acting on its outer surface. Because the engine was devised to pump water from mines, it took the form of a long pivoting beam with pump rod attached to one end and a piston rod to the other. A large piston (five to six feet in diameter) was fitted into a cylinder that had inlets for steam and the cold water used to condense the steam, and with an outlet for the waste water. After the pressure of the atmosphere had pushed the piston down to its lowest position, and lifted the pump rod to its maximum height, the weight of the pump mechanism caused the pump end of the beam to descend, raising the piston and making it possible to fill the cylinder with steam again so that the cycle could be repeated (Figure II.4). Two aspects of this engine deserve special attention: first, the weight of the atmosphere, not the expansive power of steam, did the work; and second, the cylinder was alternately heated and cooled as steam and cold water were injected into it.

In the winter of 1763/4, when Watt began his repair and study of a model Newcomen engine, the larger versions were an established power source in half the world. Despite its widespread use, some

A

B

of the features of the Newcomen engine troubled Watt and in attempting to remedy them he produced a machine that supplanted Newcomen's and prepared the way for the modern steam engine.

Watt realized that the efficiency of Newcomen's engine could be increased if the cylinder were kept uniformly hot instead of heating and cooling it during each cycle. This he accomplished by insulating the cylinder and then condensing the steam in an adjoining container that was kept constantly cool for just that purpose. In addition he abandoned the use of atmospheric pressure and moved the piston first in one direction and then in the opposite direction, by applying steam first to one side and then to the other side of the piston. Expanding steam pushing against the piston did work in a Watt engine. Thus was born the double-acting steam engine with a separate condenser. It first appeared in 1784 and dominated steam engine design for the next fifty years.

In exchanging Newcomen for Watt as the inventor of the steam engine, the issue of continuity has not been resolved; the temporal focus of the investigation has merely been changed. The question now becomes: Did the Newcomen engine appear on the scene without any antecedents? Again, the answer is no. Some of the mechanical elements that made up the Newcomen engine can be traced back to early seventeenth-century Europe, others had their origin in thirteenth-century China, and still others first appeared a century or two before the birth of Christ.

Because a Newcomen engine is mechanically more complex than a cotton gin, its antecedents are more difficult to trace in a succinct fashion. Evacuated chambers, piston pumps, steam displacement

---

Figure II.2. A. An Indian *charka*, or roller gin. The gin consists of two teak wood rollers that are rotated by turning the handle on the upper left side of the machine. When uncleaned cotton is fed between the moving rollers, the fibers pass through to the other side of the gin while the seeds remain behind and fall to the enclosure at the base. B. Eli Whitney's cotton gin opened for inspection. In operation the hinged top of the gin is closed, aligning the slotted breastwork of the top portion with the protruding wire teeth fixed to the large rotating cylinder. Rotation of the cylinder brings the uncleaned cotton to the breastwork slots, which are too narrow to permit passage of the seeds. Thus the seeds are torn away from the fiber and the fiber accumulates on the toothed cylinder. Another rotating cylinder, not shown here, is covered with bristles that brush the cleaned fiber from the large cylinder. Both the *charka* and Whitney's gin rely upon two rotating cylinders activated by a hand crank. Sources: A. Edward Baines, *History of the cotton manufacture in Great Britain* (London, 1835), p. 66. B. Mitchell Wilson, *American science and invention* (New York, 1954; copyright renewed © 1982 by Stella Adler, Victoria Wilson, and Erica Spellman), p. 80.

Figure II.3. Allegory on the significance of steam power, ca. 1850. The steam escaping from the boiling tea kettle inspires James Watt to invent the steam engine as well as envision its role in the creation of industrial civilization. This drawing is an excellent representation of the popular view that great inventions are the result of inspired intuitive leaps made by heroic figures. Source: Wolfgang Schivelbusch, *The railway journey* (Oxford, 1980), p. 5; Picture Collection, The Branch Libraries, The New York Public Library.

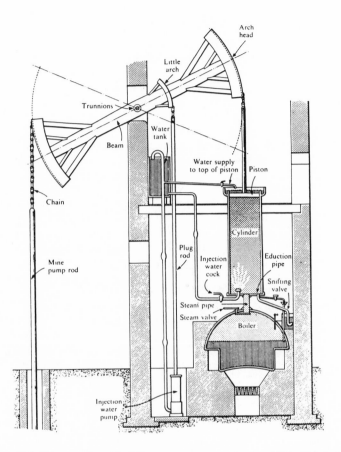

Figure II.4. Diagram of a typical Newcomen steam engine, ca. 1715. The steam-filled cylinder is about to be cooled by the injection of a spray of cold water into its interior. As a result of this action, the steam in the cylinder condenses, creating a partial vacuum. The weight of the atmosphere pressing on the piston's outer surface then forces the piston down for the engine's power stroke. When the piston reaches the lowest point on its path of travel, steam is injected into the cylinder equalizing the pressure on either side of the piston. The weight of the mine pump's mechanism then causes the beam to rotate lifting the piston to the top of the cylinder. Note that the valves and cocks that control the entry of the steam and cold water into the cylinder and the exit of the waste water through the eduction pipe are all hand-operated on this machine. Source: D. B. Barton, *The Cornish beam engine* (Bath, 1969), p. 17.

devices, and mechanical linkages all have their place in the prehistory of the steam engine. They form the "long chain of direct genetic connections" that historian Joseph Needham mapped in an essay entitled: "The Pre-Natal History of the Steam Engine." After assessing the contributions made by ancient Chinese artisans, Hellenistic mechanicians, and European natural philosophers, instrument makers, and mechanics, Needham concluded: "No single man was 'the father of the steam engine'; no single civilization either."[3] When scholars Maurice Daumas and Paul Gille investigated the background of the steam engine, they concluded that the atmospheric engine would probably have been invented in the first half of the eighteenth century even if Newcomen had never lived.

Like Whitney's cotton gin, Watt's steam engine was a seminal invention that spawned a manifold and divergent series of machines. The hot air and internal combustion engine are two of the most important power sources that have evolved from the steam engine. As early as 1759 hot air was proposed as a substitute for steam in an engine, but the first working model of such a device was not built until 1807. Later in the nineteenth century Robert Stirling, in England, and John Ericsson, in America, designed hot air engines that were sold to the public. By 1900 the hot air engine was supplanted by yet another variant of the steam engine, one that replaced the external combustion of the steam or hot air engine with an internal combustion within the cylinder. In 1791 an internal combustion pumping engine that ran on vaporized turpentine had been patented in England; however, the world's first production model of an internal combustion engine was designed by Belgian inventor Jean Joseph Etienne Lenoir in 1860. The Lenoir engine, fueled by illuminating gas, was closely patterned after a double-acting horizontal steam engine. Just as Watt's double-acting engine admitted steam on both sides of the piston and hence did work in both directions, Lenoir's engine exploded a gas and air mixture at both ends of the cylinder driving the piston forward and backward. Later improvements of the gas engine included Nikolaus Otto's single-acting four-stroke model of 1876, which served as the prototype of the modern automobile engine. Although the gaseous medium had been changed from steam to hot air to exploding mixtures of fuel and air, the basic configuration of cylinder and piston remained a constant.

The Electric Motor

Neither the cotton gin nor the engines we have considered were the immediate result of a major breakthrough in science. Therefore, one might ask, does technological change occur in a different fashion when it draws upon a recent scientific discovery? Perhaps a revolutionary development in science evokes a similar discontinuity in technology whenever it is applied in practice? To test this possibility, consider the discovery of electromagnetism by Hans Christian Oersted and its application in the earliest electric motors.

Oersted's announcement in 1820 that a current-carrying conductor produces a magnetic effect in its immediate vicinity created widespread interest throughout the scientific community. Startling as Oersted's discovery was to the world of science, its technological application followed a predictable course. The first electric motors were modeled after two well-known devices: the magnetic compass and the steam engine.

The Danish scientist had shown that a piece of wire carrying an electric current exerted a force on a compass needle causing it to deflect. An English physicist, Michael Faraday, having learned of this, immediately attempted to change the needle's deflection into continuous rotation. The result was the first electric motor. Granted, he achieved continuous rotary motion in a simple laboratory apparatus and not in a device capable of doing useful work; nevertheless, the principle of the modern electric motor had been elicited and demonstrated by Faraday. The needle of Faraday's modified compass spun continuously instead of aligning itself with the earth's magnetic field.

In 1831, within a decade of Faraday's experiments, American physicist Joseph Henry built an electric motor that drew upon steam engine mechanisms. A central feature of the Newcomen and Watt engines was a long, pivoting beam with piston connected to one end and pump rod or flywheel to the other, and Henry's new oscillating-beam electric motor likewise had a pivoting, elongated electromagnet that seesawed up and down making and breaking electric contact.

There was no electrical analogue for a cylinder and piston in Henry's motor, but several other inventors incorporated the cylinder and piston mechanism into their reciprocating electric motors (Figure II.5). Charles G. Page (1838), improving on Henry's design, used the beam as a mechanical element and converted the electromagnets into "cylinders" by shaping them as hollow coils of wire into which iron core "pistons" plunged when the "cylinders" were energized. A European inventor devised a motor with electromagnetic "cylinders and piston," beam crank, flywheel, connecting rod, eccentric valve gear, and slide rod and valves. With all of these traditional steam engine mechanisms included in the motor, only

A

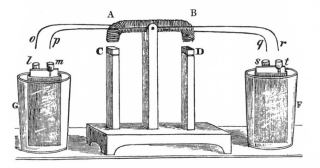

B

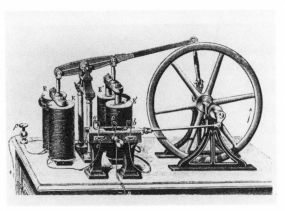

C

a boiler and firebox were needed to be added to make the analogy complete.

The natural forces that were harnessed to drive the steam engine and the electric motor were radically different ones. The scientific theories underlying the operation of these power sources – thermodynamics and electromagnetic theory – were worlds apart. The technical communities, from inventors to entrepreneurs, concerned with the development and manufacture of steam engines and electric motors were also distinct; steam engine builders did not convert their plants to electric motor manufacturing establishments. The practical uses of steam engines and electric motors were often, but not always, different. The portability of the electric motor could not be matched by the steam engine; conversely, the steam engine was far more powerful than was the electric motor. Finally, the social and economic effects of the steam engine and the electric motor were dissimilar. Even after admitting all of this, it is abundantly evident that at the artifactual level continuity prevailed; the design of the first electric motors owed more to preexisting artifacts than to scientific theory. Electromagnetic theory may have placed constraints on motor design but it did not decree that the first electric motor must operate like a steam engine.

The Transistor

At first glance the transistor, a device said to epitomize the new electronic age, also appears to be an ideal choice for partisans of the

---

Figure II.5. A. James Watt's rotative beam engine (1788). B. Joseph Henry's beam electromagnetic motor (1831). C. A rotative beam electric motor of the early nineteenth century. These three offer a striking illustration of the continuity that exists between steam engines and the first electric motors. The key mechanical element in each instance is a pivoting beam. Newcomen introduced the beam in his reciprocating atmospheric engine to operate a mine pump (Fig. II.4); Watt modified the beam mechanism to produce rotary motion in his rotative beam engine. Joseph Henry's motor, with its beam seesawing as the electromagnets A and B are alternatively attracted to C and D, recalls the reciprocating motion of the Newcomen engine while the rotative beam electric motor, like Watt's engine of 1788, transforms the reciprocating motion of the beam to rotary motion. The rotative beam electric motor incorporates other elements of the steam engine – pistons, cylinders, connecting rod, flywheel – making it a virtual electrical analogue of Watt's engine. Sources: A. H. W. Dickinson, *Matthew Boulton* (Cambridge, 1937), pl. VII; B. W. James King, *The development of electrical technology in the 19th century: the electrochemical cell and the electromagnet* (Washington, D.C., 1962), p. 260; and C. Harold I. Sharlin, *The making of the electrical age* (New York, 1963), p. 174.

revolutionary view of technological change. The transistor was first produced in the prestigious Bell research laboratories, its invention necessitated original theoretical and experimental work in solid-state physics, and its creators (John Bardeen, Walter H. Brattain, and William Shockley) were awarded the Nobel prize in physics (1956). These facts would suggest that the transistor emerged from path-breaking scientific research, that it was not part of the stream of made things.

The case for continuity seems to be further weakened by the realization that the transistor's closest relative, in terms of its uses, is the vacuum tube. Beyond the similar functions they serve, transistors and electronic tubes are far too different to sustain the claim that the former is a variant of the latter. The transistor has neither vacuum, glass enclosure, grid, nor heated cathode. Despite this evidence, one highly regarded history of semiconductor devices claims that the transistor is one of the links in "a continuous chain of new electronic devices"[4] that dates to the nineteenth century.

The search for continuity in the development of the transistor begins in the 1870s with the work of Ferdinand Braun, a German physicist who discovered that certain crystalline substances would conduct an electric current in one direction only. At the turn of the century these substances were used in crystal rectifiers to detect electromagnetic radiation. The crystal rectifiers displaced earlier detectors of radio waves and made modern radio reception possible.

The crystal radio set, with accompanying ear phones, became the first reliable and widely available radio receiver. Its main component was a holder containing a semiconductor (silicon carbide, lead sulphide, or molybdenum sulphide) and a thin, flexible piece of wire called the "cat's whisker," both of which were elements introduced by Braun in his experiments. By moving the cat's whisker carefully over the crystal in a radio set, the user could locate sensitive spots that yielded a clear signal. The drawbacks to the system were that the proper adjustment of the cat's whisker was a matter of trial and error and the crystal set could not amplify the incoming signal; nevertheless, the crystal receiver used a semiconductor for communication purposes.

The invention of the vacuum diode and vacuum triode by John A. Fleming (1904) and Lee De Forest (1906) made the crystal set obsolete by the 1920s. The new electronic tube, itself a by-product of the manufacture of the incandescent filament light bulb, amplified the incoming radio signal, making it possible to use loud speakers. While crystal detectors were relegated to amateur radio operators and to youthful experimenters interested in learning the basics of radio reception, vacuum tube technology developed rapidly.

Although toy crystal sets disappeared from the market after World War II, developments in electronics during the years preceding the war had revived interest in crystal detectors for military purposes. During the 1930s it had become apparent that short radio wave lengths eluded vacuum tubes but not crystal detectors. The advent of radar spurred renewed interest in and research on crystals that could detect these microwaves. Point-contact rectifiers resulted from this research, and they were far more sophisticated than the ones used in the old radio receivers. The crystalline material was either germanium or silicon, and the cat's whisker probe was made of tungsten.

The step from germanium microwave detectors to the first germanium transistor was neither obvious nor easy. It involved the scientific and technological labors of teams of investigators in university and industrial laboratories across the United States. The concentrated efforts of these investigators finally resulted in a workable transistor in December 1947 at the Bell Laboratories. The results were described in an announcement the following year: "In the Transistor, two point contacts of the 'cat's whisker' or detector type, familiar to radio amateurs, are made to the semiconductor."[5]

The overall design of the point-contact transistor clearly harked back to Braun's first crystal detectors as used in radio reception. Of course, theoretical understanding of the detector's operation had improved enormously by 1947, as had materials research into the kinds of crystalline substances calculated to produce a transistor effect. In addition, the early detectors acted as rectifiers whereas the transistors acted as amplifiers. Yet after all of these differences are acknowledged, the continuity of the design of the artifact remains intact.

An effective transistor *could* be designed differently; it need not be of the point-contact type. In fact Shockley subsequently invented the junction type transistor that soon replaced the original Bell model and opened the way to modern solid-state electronics. But the first transistor was of the point-contact variety, which underscores our basic rule: Any new thing that appears in the made world is based on some object already in existence.

Although crystal detectors were the main influence on the initial design of the transistor, other forces, specifically the vacuum tube, also helped to shape it (Figure II.6). Because the transistor was perceived as a replacement for a triode in a circuit, and because vacuum tube manufacturers turned to transistor production, features peculiar to thermionic tubes were transferred to the transistor. Consequently, the idea of the integrated circuit developed slowly. Traditional electronic practice had stressed the wiring together of

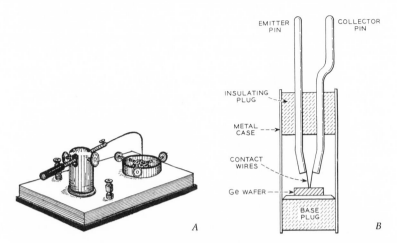

Figure II.6. A. Early twentieth-century crystal detector for use in radio receiver. The metal "cat's whisker" is mounted so that it can be easily moved to touch any portion of the surface of the crystal; the crystal is held securely in place by three set-screws. B. Schematic drawing of a point-contact transistor (1959). In the transistor the "cat's whisker" has evolved into the "contact wires" that touch the surface of the germanium wafer, the crystal's equivalent, at two points. Diagram references to "emitter" and "collector" pins are holdovers from vacuum tube terminology and inappropriate for transistors. Sources: A. Vivian J. Phillips, *Early radio wave detectors* (London, 1980), p. 207; B. John N. Shive, *The properties, physics, and design of semi-conductor devices* (New York, 1959), p. 177.

distinct components on a panel, rather than the incorporation of several components into a single manufactured unit, which is what the integrated circuit required. Likewise, the example of vacuum tube technology persuaded transistor manufacturers to treat semiconductors as if they were miniature, solid-state tubes. Borrowing from tube nomenclature and operation, the two transistor contact points and their contiguous electrical connections were respectively designated *emitter* and *collector*, even though neither emitting nor collecting was taking place. Similar analogical thinking influenced transistor manufacturers to seal their product hermetically in a glass capsule or encase it in a metal canister.

Edison's Lighting System

Large technological systems, like the discrete artifacts we have been considering, also exhibit continuity in change. Of interest here is

Thomas Edison's observation that because all components of a system must be compatible with one another then a system is like a large machine. If a steam engine with its many different mechanisms undergoes evolutionary change, should not a system made up of integrated components do likewise?

In 1878 when Edison began concentrating on a project to create an electric lighting system, there already existed two workable but quite different lighting systems in European and American cities. One used illuminating gas that was generated at a central gas works. It was piped beneath the city streets to homes, stores, and hotels, and at each of these sites tubing carried the gas to individually controlled lighting fixtures in rooms, hallways, and the like. The other system used an electric arc for lighting, in which illumination was produced when two carbon rods connected in electrical circuit were brought in close proximity. The result, an intense white light of one thousand candlepower, was useful for lighting public places; such as streets, factories, ballrooms, theaters, and auditoriums. The generating plant producing the electricity for the arc lights was located on the premises and was owned and operated by the consumer of the light. Individual arc lights were series-connected, which meant that they must all be switched on or off simultaneously. Given the public uses of arc lighting, this arrangement was not necessarily an inconvenience.

Both systems had their drawbacks. Gas lighting burned a dangerous fuel indoors, polluting the immediate environment with the products of its combustion, and it yielded a weak yellowish light of sixteen candlepower (roughly equivalent to a modern twelve-watt bulb). Arc lights flickered as their carbon rods were consumed in the intense heat, and they too released noxious fumes. Whereas the gas light was too weak for many purposes, the arc light was too strong for the ordinary home or office.

These systems were in place when Edison resolved to create one based on an electric incandescent lamp, one with a glowing filament, that would illuminate domestic and commercial interiors. His task, as he described it, was to subdivide the electric light system so that it could serve private and semiprivate purposes, not just public ones. Given his goal, one might assume that Edison would concentrate on changes in electric arc technology; instead, he decided to produce an electrical analogue of the gas lighting system.

Edison was well aware of the limited usefulness of arc lights for interior illumination. In seeking a guide for his new electrical enterprise, he chose the gas industry, which derived 90 percent of its revenue from the lighting of interior spaces. This choice led to the establishment of the first commercial electric lighting station, opened in 1882 by Edison at Pearl Street in New York City, and to the imposition of a gas lighting system model on electrical illumination systems throughout the world.

Edison's notebooks contain his earliest thoughts on the subject: "Object, Edison to effect exact imitation of all done by gas so as to replace lighting by gas by lighting by electricity . . . not to make a large light or a blinding light but a small light having the mildness of gas."[6] For Edison the electric light was not simply a new kind of lamp but a system in which the lamp was a single component to be integrated with other components: generators, conducting networks, meters, fixtures, switches, fuses, and lighting accessories. Analogues of many of these components could be found in the gas industry's successful domestic lighting system.

The hub of Edison's system was a central station, remote from the consumers of light, that provided power to dwellings and commercial establishments in a section of the city. Just as gas flowed from a central facility in large pipes, or mains, Edison chose to have electricity flow from a central generating station through copper wires. Telephone, telegraph, arc light, and fire alarm wires were all strung onto poles above ground, but Edison put his electric mains underground, explaining, "Why, you don't lift water pipes and gas pipes up on stilts."[7] Yet to gain the legal right to put his mains beneath the pavement, Edison was forced to incorporate his Edison Electric Illuminating Company under the statutes regulating the gas industry in New York State. Only gas companies were allowed to dig up the streets of the city.

There were many similarities. In Edison's electric system, as in the gas lighting system, the lights were wired so that individual lights could be turned on or off independently of the others. Just as gas meters were installed at each residence, Edison demanded residential electric meters for his system. This demand was made at a time when there was no cheap, reliable means for measuring the amount of electricity consumed over an extended period of time. Owner-operated arc light systems were self-contained and had not been metered. Only gas, and in some locales water, had been metered previously. Edison also drew comparisons between electrical and gas pressure as he explained the resistance to flow encountered in wires and pipes. Initially his electric light bulb was called a *burner* after the gas burner, and it was designed to produce sixteen candlepower as did a single gas burner. Most important, Edison carefully calculated the costs involved in the construction, operation, and maintenance of the system so that electric lighting could compete with gas lighting.

Alternative electric lighting schemes were proposed during the years Edison was at work on his system, but they never advanced beyond the theoretical stage. For example, in 1882 a well-known English electrical engineer offered a plan for domestic lighting based on the electric arc light model. Each home would generate its own electricity with a dynamo driven by a gas-powered motor. Another, more elaborate, scheme called for the installation of storage batteries in each dwelling. A central generating station would charge the batteries at high voltage during the day. At night the station would supply power for street lighting while the household batteries would provide power at low voltage for domestic illumination. Edison rejected the battery plan for storing high voltage, likening it to a gas system, proposed some time earlier, in which gas was pumped at high pressure through inexpensive, small diameter pipes and stored in a reservoir in the home where it was used at a lower pressure. He believed that high voltage posed as many dangers for the private home as did the storage of gas under high pressure.

Edison operated within the constraints of physical possibility as he worked out the details of his project; however, scientific laws alone did not dictate the overall design of his completed system. In selecting and assembling the components of his system, Edison constantly kept the technology and economics of gas lighting in mind.

To modern observers Edison's system appears to be the obvious way to do the job. The historian knows that this was not always the case. Edison's solution was certainly not obvious to his contemporaries who were well versed in the scientific and technical aspects of electrical lighting. They thought, at best, that he was engaged in a wrong-headed effort and, at worst, that he was either a fool in pursuit of the impossible or a fraud.

It is a tribute to Edison's genius that, as he developed his electrification scheme, he had the boldness and the imagination to draw analogies between two such disparate technologies as illuminating gas and electricity. That he should have felt the necessity to seek such analogies, moreover, is additional evidence supporting the model of continuous technological change. The Edison lighting example has shown that each new technological system emerges from an antecedent system, just as each new discrete artifact emerges from antecedent artifacts.

Barbed Wire

Granted that every new artifact is based to some degree upon a related existing artifact, we must next face the question of the origin of the first made thing. On what was it patterned? Although there were no preexisting artifacts at that time, a host of *naturfacts* could serve as models to initiate the process of technological evolution. There were rocks, stones, pebbles, sticks, twigs, branches, leaves, shells, bones, horns, and myriad other natural objects whose weight, structure, texture, form, and material made them suitable as found tools for the job at hand. This historical reconstruction is, of course, speculative, but it is supported by evidence unearthed by archaeologists and prehistorians who study the material culture of our most remote ancestors. They have found naturally shaped stones that show signs of wear indicating that they were used as tools by early hominids. Continuity of form between the found tool and the deliberately shaped tool is so strong that, in many cases, separating an artifact from a naturfact is difficult. The point to keep in mind is that stone tools did not appear on the scene abruptly but emerged slowly.

The transition from the first naturfacts to the first artifacts is lost in the mists of prehistoric times. We can speculate about that process but not document it in detail. There are, however, more recent artifacts whose development from a naturfact can be traced. One is an archetypical nineteenth-century American invention − barbed wire.

With the exception of stone implements, barbed wire is the simplest artifact to be discussed in this chapter. It consists of several long strands of intertwined wire onto which shorter pieces are wrapped perpendicularly at regular intervals. The exposed ends of the shorter lengths are clipped at an angle to transform them into sharp points or barbs. When the wire is strung on posts the barbs project on both sides of the fence acting as a deterrent to captive cattle who might want to roam beyond the boundaries of the property or free-ranging livestock who try to feed on crops protected within the fence. It is a simple, cheap, and very effective barrier.

The ease with which barbed wire can be made has led historian of technology D. S. L. Cardwell to suggest that it could well have been invented long before the third quarter of the nineteenth century, perhaps in ancient Greece. We might question the idea of barbed wire in classical antiquity and move the date forward to the Renaissance when wiredrawing was first practiced on a large scale, but even that shift would not invalidate Cardwell's contention that such a simple artifact, crafted from twisted lengths of wire, could have been made much earlier than it was. The invention of barbed wire certainly did not depend on the advancement of scientific knowledge or on the perfection of some complex and precise technological process. Why then did it first appear in late nineteenth-century America? Or more specifically, what were the prevailing conditions that led three men to invent barbed wire in DeKalb, Illinois, in 1873?

The first settlers in America brought with them traditional English and European ideas on how to build a fence for the enclosure of their agricultural plots. In most cases these fences were constructed of stone or wood, two materials readily available in the early colonies. So long as settlement and agriculture were confined to the Atlantic seaboard, fencing proved to be no problem, but in the nineteenth century the nation expanded into the western prairies and plains. Migrating farmers found that wood was scarce and expensive there and that their crops needed protection from the cattlemen's herds that ranged freely seeking food. Thus, fencing quickly became the farmer's primary concern. Between 1870 and 1880 newspapers in the region devoted more space to fencing matters than to political, military, or economic issues.

The prohibitive costs of wooden fences slowed the westward expansion considerably. In 1871 the U.S. Department of Agriculture calculated that the combined total cost of fences in the country equaled the national debt, and the annual repair bill for their maintenance exceeded the sum of all federal, state, and local taxes. Because of a desperate need for reasonably priced alternative fences, several new types were tried.

One of the most successful alternatives, well known in Europe but not widely used heretofore in America, was the hedge row. In a region where free-ranging cattle threatened crops, the effectiveness of hedge rows was increased when thorn-bearing plants were used to form them. Briar, mesquite, cactus, rose, and various locusts were grown but Osage orange appeared to be the best plant for fencing purposes.

Osage orange, or *bois d'arc* as the early French traders called it, is a rather short tree with pronounced thorns on its branches. It can be grown as a tall shrub, and if planted close together in double rows, and pruned to encourage growth on its lower parts, Osage orange in three or four years becomes a "living fence" that resists the incursion of cows, horses, and pigs. Because Osage orange was native to eastern Texas and the southern parts of Arkansas and Oklahoma, but could be grown in colder climates, its cuttings were propagated in Texas and Arkansas and its seeds processed there for shipment northward to the prairie states. During the 1860s and 1870s, Osage orange culture became a thriving, though modest, industry. In 1860 alone 10,000 bushels of seed were sent north. This was sufficient to produce 300 million plants or about 60,000 miles of hedge.

For a time, a hedge of thorny, impenetrable Osage orange

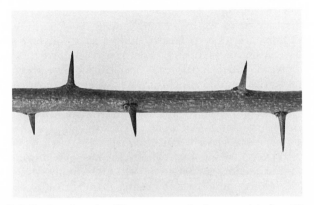

Figure II.7. Section of a twig of Osage orange, also known as *bois d'arc*. The thorns are about one-half inches long and are set about one and one-half inches apart. Barbs on a strand of modern barbed wire are one-half inches long and are positioned at five-inch intervals.

was thought to be the solution to the fencing problem. But hedges, even those as tough and useful as Osage orange, generated their own sets of difficulties. They were slow to develop, could not be moved easily, cast shadows on adjoining crops, usurped valuable growing space, and provided a shelter for weeds, vermin, and insects.

Whatever its drawbacks the thorny hedge, and especially *bois d'arc*, was the naturfact that served as the model for the immensely successful barbed wire that was soon to fence the West (Figure II.7). More than a "living fence," Osage orange is "living barbed wire." The truth of this claim can best be appreciated by observing a branch of Osage orange closely. The elongated sturdy thorns, all of equal length, are attached perpendicularly to the branch at regular intervals and encircle it uniformly in a spiral pattern. Thus, the design and mechanical regularity of factory-made barbed wire was foreshadowed in the natural form of a branch of Osage orange.

In addition to hedge rows, smooth wire fencing had been used in timber-scarce areas of the West. Wire was cheap to purchase, transport, and install; it produced no shade and harbored no pests; and it could be easily shifted to new boundary lines. Unfortunately, it offered little hindrance to cattle who routinely broke through it even when multiple strands were affixed to the posts. The ultimate fence, it seemed, would be one that combined the best features of wire and "living" fences.

Such a combination was apparently what Michael Kelly had in mind in 1868 when he patented an improved type of fencing (patent no. 74,379). "My invention," he wrote, "[imparts] to fences of wire a character approximating to that of a thorn-hedge. I prefer to designate the fence so produced as a thorny fence."[8] Kelly used a single strand of wire fitted with diamond-shaped, sheet metal "thorns" at six-inch intervals. The Thorn Wire Hedge Company was founded in 1876 to manufacture Kelly's invention but by that time barbed wire from DeKalb, Illinois, had come to dominate the fencing scene.

Kelly's fence was one of several dozen examples of fences armored with spikes or barbs that had been invented, and in some cases patented, between 1840 and 1870. None of these earlier efforts yielded a commercial product that was used on farms prior to 1873. Hence the honor of inventing barbed wire is reserved for the three citizens of DeKalb who devised, produced, and sold the first substantial quantities of barbed-wire fencing.

DeKalb, Illinois, located at the edge of the prairie, was a likely place to find men interested in designing new fences. The farmers and mechanics of DeKalb were well aware of the need for inexpensive, effective fencing in the broad treeless regions of the West and were ready to take advantage of any innovation that promised to meet that need.

At the DeKalb County fair in 1873 Henry M. Rose exhibited a device that could be attached to existing plain wire fences to deter cattle from breaking through them. The attachment was a piece of wood, one inch square and sixteen feet long, into which long brads were driven so that their sharp points protruded along its surface (patent no. 138,763). Rose's fence attachment caught the attention of three men who had come to the fair: Jacob Haish, a German-born lumberman, Isaac L. Ellwood, a hardware merchant, and Joseph F. Glidden, a farmer. Each of them left Rose's exhibit convinced that he could create a superior fence by making the barbs an integral part of the fence wire. They succeeded in doing so and originated the large-scale manufacturing of barbed wire with Glidden and Ellwood joining forces in one factory and Haish founding a rival establishment (Figure II.8).

In 1874 the infant barbed-wire industry produced 10,000 pounds of the new fencing. The response to barbed wire was so favorable that within a few years it was being shipped out of the factories in railroad-car lots: 600,000 pounds in 1875; 12,863,000 pounds in 1877; and 80,500,000 pounds in 1880.

The DeKalb inventors were all acquainted with hedge fencing

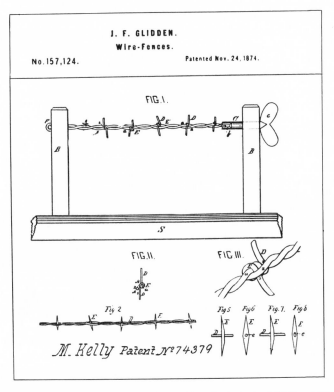

Figure II.8. Patent drawings of Joseph F. Glidden's barbed wire (1874) and Michael Kelly's thorny fence (1868). Kelly's fencing, illustrated at the bottom, consists of two strands of wire twisted together to form a cable on which are strung flat metal thorns (Fig. 6). Glidden's wire, one of the most popular barbed-wire types ever manufactured, features a two-strand twisted wire cable that incorporates coiled wire barbs at regular intervals (Fig. III). Fig. I. shows Glidden's wire attached to a key that can be turned to tighten the fence when it sags. Source: Henry D. McCallum and Frances T. McCallum, *The wire that fenced the West* (Norman, Okla., 1965), facing p. 81.

and Haish had a special interest in reusing *bois d'arc*. In 1881 he recalled his earlier involvement with that plant:

In the late 60s and early 70s, the planting of willow slips and osage orange seed was at a fever heat. I had received a consignment of osage orange seed from Texas, supplying some of my customers with the same at $5.00 per pound. . . . It was in my mind [at one time] to plant osage orange seed and when of suitable growth cut and weave it into plain wire and board fences, using the thorns as a safeguard against the encroachment of stock.[9]

Barbed wire was an extraordinary invention that revolutionized fencing in America and other parts of the world, facilitated the westward movement, brought the fruits of the Industrial Revolution to the farm, and had a profound effect on the farming and cattle industries as well as on warfare and prisons. Although simple in conception, it appeared at a late date in the history of made things. It is a modern example of the process by which a naturfact is transformed into an artifact and shows that even the simplest of artifacts has an antecedent. Barbed wire was not created by men who happened to twist and cut wire in a peculiar fashion. It originated in a deliberate attempt to copy an organic form that functioned effectively as a deterrent to livestock.

A Book-Writing Machine

If the artifacts of material culture emerge from other artifacts, or in some special cases from naturfacts, from whence come the imaginary machines we encounter in the fantastic worlds created by writers and artists? Are these pure products of the artistic imagination or can they be placed in the stream of made things we have been exploring? The final test of the continuity thesis will be an attempt to answer these questions by taking a close look at a fantastic machine, one that has not, and probably could not, be fabricated in the real world.

The heroine of George Orwell's *1984* is a young woman who works in the Fiction Department of the Ministry of Truth. Carrying a wrench in her oil-stained hands, she maintains the novel-writing machines that churn out popular stories for the masses. Orwell does not describe this wonderful machine in any detail but Jonathan Swift, from whose *Gulliver's Travels* (1726) Orwell borrowed the idea, offered a verbal description of the mechanism, outlined its operation, and provided readers with a picture of it.

Captain Lemuel Gulliver's journeys took him to the country of Lagado, where he was given a tour of its academy of the arts and sciences. There, in the speculative learning section, he was shown a large machine capable of writing books on any subject. With little intelligence, skill, or knowledge, an operator could write books in philosophy, literature, politics, law, mathematics, and theology. The machine, in the form of a twenty-foot square, was composed of wooden cubes (or "dies" as Swift called them) onto whose surfaces were glued slips of paper. These slips of paper held the entire vocabulary of the Lagado language, one word to each side of a cube. The cubes were uniformly attached to one another by wires or rods in such a way that they displayed a random combination of words whenever they were rotated. Rotation was achieved by means of iron handles, or cranks, that projected from the edges of the machine's frame. The illustration that accompanies Swift's text depicts a device that is mechanically inoperable but one that is a fair approximation (although with fewer cubes and handles) of what the author had described.

To write a book, the "author" stationed an assistant at each of the cranks. At a signal from the "author," the forty assistants simultaneously rotated the cubes through at least ninety degrees thus exposing a new set of words on the surface of the machine. Thirty-six of the assistants then scanned this scrambled assortment of words searching for any meaningful phrases that might have appeared by chance. These phrases were copied down by the remaining assistants, the cranks were turned once again, and the process repeated until the broken sentences or phrases filled several folios. From these the author would write a book. The inventor of this device, certain that the mechanization of the writing of books would enrich humanity, was seeking public funds with which to build five hundred machines for use throughout the country.

Swift's discussion of the Lagado machine was clearly satirical. He was ridiculing the idea that chance or accident governed intellectual creativity and that machines could duplicate the actions of the human mind. Still, there remains the unresolved problem of the origins of Swift's fanciful contrivance.

Some literary historians have linked the book-writing machine to the mechanical mathematical calculators built in the seventeenth century. However, these calculators, with their gears, calibrated dials, and recording hands, are closely related to clockwork mechanisms and far removed from Swift's cumbersome rotating cubes. But it is true that some of the early calculators were activated by small metal cranks. Other scholars have suggested that Swift, bent on satirizing contemporary attempts to create a universal language, merely added a set of cranks to the printed tables of words that commonly appeared in the books written by promoters of a universal medium of discourse. This second explanation of the origin of Swift's machine is unconvincing on several counts. The tables were presented in two, not three, dimensions, and there is no indication that the words printed in the universal language tables were ever meant to be separated from one another and affixed to other surfaces, and rearranged. There are, however, obvious artifactual antecedents for Swift's machine that have been overlooked – namely, children's alphabet blocks, which can be manipulated to spell out simple words (Figure II.9).

Cubical blocks of wood or ivory with different letters of the alphabet engraved on their faces were used in Elizabethan England to teach children their ABCs. In the seventeenth century, these educational toys gained new prominence when philosopher John Locke recommended that "Dice and play-things with . . . Letters on them"[10] be given to young children so that they might learn the alphabet while they played. By the eighteenth century entire words written on pieces of paper, and pasted to slips of wood or to "dies," were used to teach spelling. These educational artifacts form the background for Swift's invention of the book-writing machine.

In his mind's eye, Swift took alphabet blocks, glued words to their surfaces, and joined them mechanically so that they might generate phrases randomly. Just as the child arranged letter cubes to spell words, the machine arranged word cubes to form fragments of sentences. The alphabet blocks not only provided a suitable mechanical element for Swift's device, they also contributed to the satire by underscoring the childishness of the Lagado invention.

Even in this exercise of the literary imagination, Jonathan Swift was influenced by things that already existed in the material world. And what was true of the book-writing machine is also true of all imaginary contrivances, be they creations of science-fiction authors or the technological dreams of engineers and inventors. Like their counterparts in the real world, fantastic machines follow the rule of continuous development.

The Origins of the Discontinuous Argument

Despite evidence to the contrary, there is widespread support for the idea that inventions are the result of revolutionary upheavals in technology brought about by individual geniuses. The sources of this outlook are threefold: the loss or concealment of crucial antecedents; the emergence of the inventor as hero; and the confusion of technological and socioeconomic change.

Given the nature of technology and technological change, inventor and public alike are apt to forget, or at times deliberately suppress, the debt owed to a key antecedent. Whitney's first gin bore a strong resemblance to the Indian *charka*, but that likeness was quickly lost as the machine evolved into its modern form. Few realize that important features of the modern automobile's form, structure, and mode of manufacture were derived from the bicycle, yet the first automobiles were little more than four-wheeled bicycles – Henry Ford called his invention a quadracycle – powered by gasoline en-

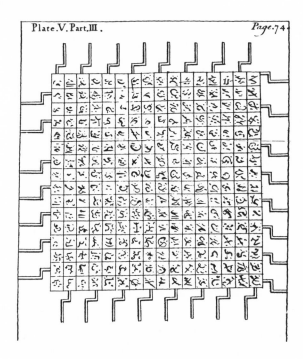

Plate.V.Part.III.   Page.74.

A

44. *A ready way for children to learn their A.B.C.*

CAuse 4 large dice of bone or wood to be made, and upon every square, one of the smal letters of the crofs row to be graven, but in some bigger shape, and the child using to play much with them, and being always told what letter chanceth, will soon gain his Alphabet, as it were by the way of sport or pastime. I have heard of a pair of cards, whereon most of the principall Grammer rules have been printed, and the School-Master hath found good sport thereat with his schollers.

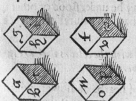

B

gines. Likewise, few are aware that modern electronic digital watches share with the oldest mechanical clocks a peculiar mode of measuring time – the division of a temporal interval into equal discrete units or beats. An escapement performs this function in a mechanical clock, a vibrating quartz crystal does it in the digital watch. Although this division into beats is not the only way of telling time, it was the first successful way of doing so and it has persisted.

Antecedent loss or concealment has occurred throughout history; the creation of the myth of the heroic inventor, however, is confined mainly to the past 300 years. Before the eighteenth century, inventors did not routinely gain especial recognition for their contributions. The history of earlier technology is largely an anonymous one with a few prominent names remembered.

The period of vast social and economic changes we call the Industrial Revolution brought numerous inventors to public notice and acclaim. They were granted recognition for having conceived the ingenious machines that fostered current progress in the economic, social, and cultural realms. Raised to the status of a military or political leader, the nineteenth-century inventor was presented as a romantic hero who battled social inertia and confronted powerful natural forces in order to bestow the gifts of technology upon humankind.

During this age, Samuel Smiles wrote books celebrating the lives and endeavors of British engineers thereby establishing a widely copied literary genre – the popular biography of the inventor. The first international industrial exhibitions, beginning with the Crystal Palace exhibition of 1851, put the machine and its products on display for the instruction and amusement of the general public. A contemporary writer in the American *Christian Examiner* (1869) could state that while all poets, philosophers, and theologians had a strong tendency to be "mean and little," all inventors were "heroic and grand."[11] Clearly these were not the times to detract from the singularity of the inventor's feat. Because heroic deeds are most often

---

Figure II.9. A. The Lagoda book-writing machine. Each square on the illustration represents a cube whose surfaces are covered with words. A turn of the crank exposes a new assortment of words that might contain a meaningful phrase or two. Notice that the cubes are expected to rotate about two axes perpendicular to one another! Swift offered no explanation for the resolution of this mechanical problem. B. Children's alphabetic blocks. Widely known in Elizabethan England, they are likely antecedents for the cubes on the Lagoda machine. Sources: A. Lemuel Gulliver, [Jonathan Swift] *Travels into several remote nations of the world* (London, 1726), p. 74; B. Hugh Plat, *The jewel house of art and nature* (London, 1653), p. 42.

linked with revolutions, evolutionary explanations of technological change did not have a broad appeal.

Nationalism also played a part in maintaining the nineteenth-century belief that technological development was essentially discontinuous. The same exhibitions that glorified industrial progress, and the men who made it possible, were also used to measure the relative industrial growth of nations through an award system that honored those countries with the greatest industrial accomplishments. For the first time in history, technological achievements were included in the determination of a nation's status in the world. Technology became a factor in international affairs and rivalries.

Given this amalgamation of technology with national interests and prestige, patriotic pride dictated the writing of chauvinistic histories of inventions that attributed the most important ones to fellow countrymen and passed over the work by individuals in other countries, no matter how talented or influential these inventors might be. A bizarre situation thus developed in which the heroic inventors of one country were scarcely acknowledged in another land. To take a well-known example, the "inventor" of the incandescent electric light bulb is Sir Joseph W. Swan in Britain, Thomas A. Edison in America, and A. N. Lodygin in Russia. Similarly, the Russian assertion that A. S. Popov invented radiotelegraphy is disputed by those in the West who designate Guglielmo Marconi as the inventor. In sum, parochialism limits the acknowledgment of the prior work done by technologists in other countries, focuses attention upon the *de novo* emergence of inventions from the solitary labors of heroic nationals, and favors a revolutionary approach to technological change.

The patent system is another modern development that has contributed to the support and dissemination of the discontinuous argument. Patents are the legal means by which industrial societies reward and protect technological innovators. In the process of doing so, an invention is uniquely identified with its inventor and its associations with existing artifacts is obscured. All of patent law is based on the assumption that an invention is a discrete, novel entity that can be assigned to the individual who is determined by the courts to be its legitimate creator. Thus, the patent system converts the continuous stream of made things into a series of distinct entities.

In a capitalistic society, the holder of a patent is in a position to use the patent for personal financial advantage. Because money, social status, and ego gratification are all at stake, the contenders in a patent dispute often fight less than fairly to preserve their claim to originality. Samuel F. B. Morse, for example, stoutly and falsely denied that he had ever learned anything crucial to the development of the electric telegraph from physicist Joseph Henry. Eli Whitney, in the midst of securing a patent for his cotton gin, asserted that he had never seen the improved roller gins that had been devised to attempt to clean short staple cotton. (He did not declare, however, that he had never encountered the older *charka* gins that had undoubtedly influenced him.) Even Thomas A. Edison was not above making dubious claims when he sought recognition for the invention of moving-picture apparatus. Such dissimulations are the result of a system that attempts to impose discontinuity on what is essentially a continuous phenomenon.

In granting a patent the government does more than give its originator an exclusive legal right to exploit it. A patent bestows societal recognition on an inventor and distorts the extent of the debt owed to the past by encouraging the concealment of the network of ties that lead from earlier, related artifacts.

The final source of the revolutionary explanation for technological change is the confusion of technology with its social and economic ramifications, best exemplified by the title Industrial Revolution. In the early nineteenth century, it meant a series of crucial inventions that transformed industry. The revolution was assumed to have occurred first within technology and then spread to industry. This meaning persists in modern usage in phrases such as the "Second Industrial Revolution" and "Third Industrial Revolution," referring to fundamental changes in industry caused by the introduction of electronics and computers. A second meaning, and one that has a wider currency, defines the Industrial Revolution as a major alteration in society brought about by technology. That is how Friedrich Engels used the term (1845) when he wrote that a revolution had "changed the entire structure of middle-class society" in England.[12] According to the first definition the technological–industrial change is revolutionary; according to the second it is the social and economic changes that are so. Because in current practice these two definitions have been merged, it is not always clear precisely what has undergone a revolution.

The industrial changes of the late eighteenth and early nineteenth centuries were truly revolutionary in the ways they affected the lives and fortunes of the people of Great Britain. Yet the machines, and the steam engines that powered them, were the outcome of evolutionary changes within technology. Neither marked an abrupt break with the past. The economic and social consequences of these developments, on the other hand, were so far-reaching that they transformed the social order.

Upheavals in the social and economic spheres have all too often been interpreted wrongly to signify revolutionary changes in technology. The establishment of the first industrial society in Britain was a change of such magnitude that it overwhelmed the technological continuity on which it was based and helped to perpetuate the view that technology advances by leaps from one great invention to another.

The confusion between technology and its consequences joined the myths of the heroic inventors, the ideas of material progress, nationalism, and the patent system and furthered the discontinuous explanation of technological change. Only a close study of artifacts can demonstrate the inadequacies of that outlook and the relevance of the continuous argument.

Conclusion

In assessing the wider implications of the continuity argument, we must be careful to avoid the implication that inventions are inevitable, or that the stream of made things is entirely self-generating and self-motivating. Given the *charka*, it was not foreordained that Whitney's cotton gin would appear on the scene in 1793. The social, cultural, economic, and technical forces that created the need for a better way to clean short staple cotton came together in the American South during the last decade of the eighteenth century. An alternative environment in which cotton was not a desirable textile or one in which cheap labor was plentiful would not have encouraged a search for new ginning techniques. A talented inventor and a likely antecedent are necessary, but not sufficient, conditions to create an innovation with wide social and technological repercussions. The new cotton gin did not necessarily have to be based on the *charka*. A gin working on different mechanical principles than those embodied in Whitney's machine could have been devised. Continuity requires artifactual antecedent but does not decree that only one artifact can play the role of antecedent as individuals search for a solution.

In most instances the antecedent will be one that exists within the general area of technology in which the innovation is being sought. Whitney drew upon the *charka*, and the contrivers of barbed wire were inspired by thorn-hedge fences; however, the inventors of the electric motor turned to steam power technology for guidance, and Edison reached into the field of gas illumination for his model of an electrical lighting system.

Functional requirements have always had a strong influence on the choice of an appropriate antecedent and because functionality may well cut across established technological boundary lines, the antecedent may not always be the one that appears initially to be the most obvious one. This was the case with the invention of the mechanical reaper (1780–1850).

In the earliest, and unsuccessful, mechanical reapers, attempts were made to duplicate the swinging motion of the scythe as it *cut* through the grain or to imitate the *clipping* action of scissors or shears. The McCormick reaper, which brought large-scale mechanical reaping to the farms of America, utilized an oscillating serrated (toothed) blade to *saw* through the grain stalks. McCormick's machine copied the action of the very ancient hand sickle, whose serrated blade was used in a sawing motion to sever the stalks. In each of these cases an artifact served as a model for the cutting mechanism: scythe, scissors, sickle. As it happened, the most obvious of the choices, the scythe, proved least useful in meeting the functional requirements of a mechanical reaper and the most primitive of them, the sickle, opened the way for the mechanization of harvesting.

The evidence for artifactual continuity presented in this chapter does not negate the fact that a Whitney, Watt, or McCormick was dissatisfied with existing technological solutions and in search of new ones. Here the continuous nature of that search has been stressed; in the next two chapters the psychological, intellectual, social, economic, and cultural aspects of the technologist's pursuit of novelty will be explored. If evolutionary change is to occur, then novelty must find a way to assert itself in the midst of the continuous.

CHAPTER III

Novelty (1): Psychological and Intellectual Factors

Introduction

The diversity that characterizes the material objects of any culture is proof that novelty is to be found wherever there are human beings. If this were not the case, strict imitation would be the rule, and every newly made thing would be an exact replica of some existing artifact. In such a world technology would not evolve; the range of material goods would be limited to the first few naturfacts used by the earliest men and women.

If we accept the proposition of universal artifactual diversity, we must acknowledge that a greater variety of artifacts is available in some cultures than in others. At one extreme there is the United States that currently issues about seventy thousand patents annually and at the other extreme are the Australian aborigines or the native inhabitants of the Amazon basin whose meager stores of tools and utensils have changed very slowly over many centuries.

How can we account for differences in the rate of production of new kinds of things? And how can we identify the sources of novelty in any culture? Answering these questions is no small task. The study of innovation is filled with confusing and contradictory data, theories, and speculations. Because there is no consensus on how novelty emerges in the modern Western world, we cannot expect to find reliable guidelines to an understanding of innovative activity in our own past, let alone in the histories of cultures radically different from ours.

That the process of innovation involves the interplay of psychological and socioeconomic factors is generally agreed. An overemphasis on the psychological elements leads to a genius theory of invention, one in which the contributions of a few gifted individuals are featured. An excess concentration on the social and economic elements yields a rigidly deterministic explanation that presents an invention as the inevitable product of its times. Because it is so much easier to identify socioeconomic influences than it is to delve into the workings of innovative minds, and because we have yet to produce a theory capable of fully integrating the psychological, the social, and the economic in any realm, a satisfying unified explanation of innovation remains more an ideal than a reality.

In the discussion that follows we shall attempt to maintain a balance between the internality of the psychological and the externality of the social and economic. The findings of psychological research into the wellsprings of creativity are not included, because that material is not immediately pertinent to the theory of technological evolution.

Along with a discussion of the psychological factors affecting the emergence of novelty, this chapter will also consider the role knowledge plays in technological innovation. The next chapter will focus on the social, economic, and cultural forces that encourage the search for novel solutions to technological problems. This division of topics, made for the purposes of analysis, cannot be rigidly maintained. In many instances several factors coalesce to influence the emergence of novelty.

An assumption that permeates the argument of this chapter and the next is that the potential for invention exists throughout the human race. Some individuals have greater inventive skills than others, some cultures are better able to exploit the innovative potential in their midst, and in some cultures inventiveness strongly manifests itself in other than novel material objects. But there is no evidence to support the contention that a particular nation or race has an extended monopoly on creativity.

When modern Westerners meet a people whose material culture has much less diversity than their own, they are tempted to make invidious comparisons between spear and rifle, grass hut and skyscraper, or bark canoe and airplane, and to attribute the lack of material progress to the inferiority of the primitive mind. A more reasonable explanation is that some societies have fashioned a way of life that simply does not place great value on technological change and its accompanying artifactual diversity. One such group of people is the Tikopia, Polynesian islanders who were studied by anthropologist Raymond Firth in the late 1920s.

The home island of the Tikopia contains no minerals, few stones, and no clay for building or pottery. Plant fibers, wood, and a little iron gained by trade are used in making canoes, clothing, and tools.

The Tikopia showed a definite lack of concern for technological change. Firth found that they were not particularly interested in making new things or in improving traditional techniques for making old ones. Although they acknowledged the superiority of the white man's artifacts, they were not envious of foreign technological success nor did they long to emulate it.

The Tikopia were not prohibited by religion or magic from accepting Western technology. They traded openly for metal tools, European cloth and beads, and imported food plants. Nor were they incapable of adapting alien technology to suit their purposes. They fitted the steel blades of Western carpenter planes into native adzes, borrowed a brace and bit to bore holes when constructing their canoes, and shaped the handles of discarded Western toothbrushes into earrings. In short, although the Tikopia showed evidence of inventive potential, they lacked the ambition or interest to pursue technological novelty with any vigor. Living in a well-integrated culture that rewarded conformity to established rules and ways, they had no incentive to seek technical advances. By Western standards, the Tikopia were technologically stagnant; according to their own value system, technology was in its proper place and in harmony with the rest of their culture.

Fantasy, Play, and Technology

In a conventional treatment of the development of technology, the search for novelty begins with an invocation of *homo faber* (man the maker) and the ways in which the quest for the necessities of life has inevitably led to artifactual diversity. Instead, we shall look to *homo ludens* (man the player), who will introduce the topic of novelty, and then we shall consider how the role of play serves as a source of technological innovation.

A number of writers on the topic of technological innovation have acknowledged the significance of play and have commented on the pleasures derived from the game of invention, apart from any economic or social gains it might bring. Inventors gain much satisfaction from solving the puzzles they encounter, overcoming the challenges set before them, and pitting their intellects against nature and human competitors to win the game.

An element of make-believe dominates play, so we will emphasize the role of fantasy; however, fantasy is so broad a topic that we shall divide our discussion into three parts: technological dreams, impossible machines, and popular fantasies.

Technological Dreams

Technological dreams are the machines, proposals, and visions generated by the technical community, whether in the Renaissance or the present time. They epitomize the technologists' propensity to go beyond what is technically feasible. Fanciful creations of this kind provide an entry into the richness of the imagination and into the sources of the novelty that is at the heart of Western technology. They also challenge the conventional depiction of the technologist as a rational, pragmatic, and unemotional person dominated by a utilitarian outlook.

Technological extrapolations are the first examples of the playful creations that emanate from technical minds. The majority of these are relatively conservative ventures well within the bounds of possibility, perhaps a step or so beyond the current state of technology; they normally offer no serious challenge to the status quo. Nevertheless, most of these extrapolations will probably never be constructed (Figure III. 1). For this reason they might be thought of as imaginative exercises, or as elegant variations, based upon well-known technological themes. Because these working devices and apparatuses exist for the most part as illustrations in books, they are not only the dreams of the technologists who first created them but also of all those who have subsequently enjoyed contemplating and learning from the ingenious solutions they supply.

The contents of the machine books dating from the Renaissance provide an excellent opportunity to survey the dreams of early modern technologists. Between 1400 and 1600 a number of elaborately illustrated books of this sort were published in Germany, France, and Italy. Some of them were descriptive in nature, accurately presenting current technological practices and artifacts in fields such as mining and metallurgy. But another, and very influential, group of them contained hundreds of pictures of machines that were extrapolated from existing technology. These volumes were repositories of novelties that had not yet been built but that were depicted with such care and authenticity that they might possibly be constructed in the future. *Theatrum machinarum* (theater of machines) was the title given to these books, and rightly so for they presented technology as a spectacle for the enjoyment and instruction of the reading audience.

One of the most popular of the *theatrum machinarum* was *Le diverse et artificiose machine* (*The Various and Ingenious Machines*) by Agostino Ramelli, a French military engineer. First published in 1588 Ramelli's book was reprinted, translated, and portions recopied during

Figure III.1. Hydraulic spinning wheel. The spinning wheel is powered by the water flowing from channel M into and through the tube wound spirally around cylinder K. Giovanni Branca, in whose 1629 book this machine appears, assures us that it can be used to twist, spin, or wind thread. He does not explain why so much power, and so ingenious an apparatus, are required to turn a spinning wheel that was normally activated by a single foot pedal. Source: Reprinted with permission of Macmillan Publishing Company from *A theatre of machines* by Alex G. Keller, pp. 32–3. Copyright © 1964 by Alexander G. Keller.

the next four centuries. The devices that Ramelli pictured were ordinary enough, but they were presented in such variety, and their mechanisms fashioned with such genius, that his work is far more than a textbook or manual for aspiring engineers. It is a celebration of technological possibility. Ramelli presented 110 water pumps (Figure III.2), 20 grain mills, 14 military screwjacks for use in breaking down doors and forcing iron grates, 10 cranes – and everyone of them different. As Eugene S. Ferguson, Ramelli's modern editor, wrote: "Ramelli was answering questions that had never been asked, solving problems that nobody but he, or perhaps another technologist, would have posed."[1]

Economic necessity certainly was not the motivating force behind the plethora of technological novelties. They were the products of a fertile imagination that took delight in itself and in its ability to operate within the constraints of the possible, if not the useful. Some of the novel mechanisms pictured in the machine books were later incorporated into practical devices; others stand unused as proof of the fertility of the contriving mind.

Patents comprise the second group of technological dreams. Their inclusion here calls for some explanation because patents are usually awarded for innovations that have passed the careful scrutiny of patent office examiners and are not fanciful schemes. Taken as a whole, however, patents are better representatives of technological potentiality than they are of technological actuality.

Patented devices exist in some sorts of prototypes that operate as claimed, but this fact does not mean that they will necessarily reach the marketplace. In 1869 U.S. Commissioner of Patents Samuel S. Sparks estimated that 10 percent of all patents had commercial value. Although nearly a century later economist Jacob Schmookler estimated the figure was 50 percent, many modern commentators agree with Sparks. Most patents are never commercialized but remain unused in the files of the patent office.

The technological potential of Western societies is even greater than this discussion indicates because there are probably as many innovators who have not gone to the trouble and expense of seeking patents for their inventions as there are recognized patent holders. At the very least, several hundred thousand patented and unpatented inventions are produced annually in the United States alone.

The mention of untapped innovative potential conjures up images of great inventions – on the order of the steam engine, telephone, or transistor – lying idle in the patent office or workshops of inventors. Unfortunately, this is not the case. Most inventions are of a modest if not trivial nature. They are hardly the sort to transform our technological world. In looking through the lists of inventions patented in the United States, one must search very diligently for the few familiar and important machines known to history. In most cases they are overwhelmed by quite ordinary patented novelties.

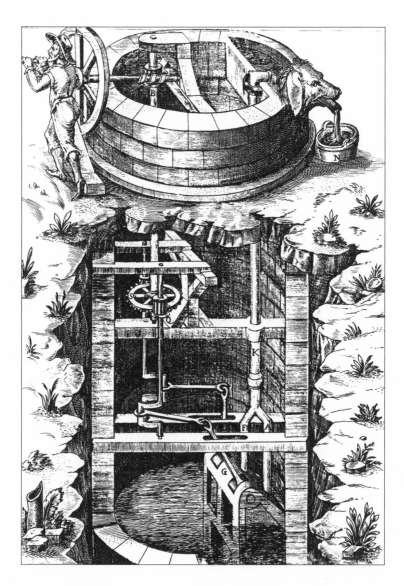

Figure III.2. A Ramelli water pump. The main pumping mechanism consists of two large paddles at the bottom right of the well. Turning the crank causes the paddles enclosed in chamber G to drive the water into pipes F and A and past a valve, in pipe K, that prevents the fluid from flowing back down into the well. The result is that water is forced up pipe K until it pours from the dog's mouth into basin N. Source: Agostino Ramelli, *The various and ingenious machines of Agostino Ramelli* (Baltimore, 1976), pl. 68.

The argument that economic incentives were the driving force behind the invention and patenting of a majority of novel artifacts is not persuasive. Although many inventors were motivated by the unrealistic belief that their particular gadget would earn them a fortune, others pursued novelty for the psychic rewards it brought. In neither instance, however, do we find inventors working to supply pressing human needs or carefully appraising economic conditions, calculating precisely what innovations are most likely to bring the higher financial returns. For this reason, many patent holders belong in the company of the technological dreamers who repeatedly, enthusiastically, and ingeniously provide solutions to problems that are mainly of concern to themselves.

Technological visions, the final category of technological dreams, are bold and fantastic schemes ranging from the improbable to the edge of the impossible. They are the means which technologists for the past five hundred years have used to express the most extravagantly fanciful aspect of their innovative activity. Yet these visions should not be confused with science fiction. As creations of the technological, not the literary or popular, imagination they are essentially an exaggerated form of the element of play found earlier in extrapolations and patents (Figure III.3).

The earliest technological visions date to the fifteenth century, when there first began to appear treatises featuring machines that were so far beyond the reach of current technology that they could not be shown in full technical detail. One of the earliest of these books was Conrad Kyeser's *Bellifortis* (1405), noted for its many fantastic war machines. In *Bellifortis*, as elsewhere in the genre, it was not expected that the devices pictured were actually to be built.

The most famous collection of Renaissance visionary machines was not revealed to the public until late in the nineteenth century. It was hidden away in the unpublished personal notebooks of Leonardo da Vinci (1452–1519). Leonardo's drawings contain some of the best examples of fanciful devices ever produced. Among them are sketches of flying machines (both powered and free flight), parachutes, armored tanks, gigantic crossbows and catapults, a small battleship, multibarreled guns, a steam engine, and a steam cannon. He also offered plans for paddle boats, diving suits, various dredging vessels, and a self-propelled spring-powered wagon. Many of these are inoperable as displayed and few, if any, influenced the subsequent growth of technology; however, they permit a rare insight into the mind of a great technical genius and into the kind of technological exuberance that was to become one of the hallmarks of Western civilization. To the best of our knowledge Leonardo's fantastical

Figure III.3. Self-propelled Renaissance vehicle. One of ten fantastic carriages, all human-powered and provided with different, elaborate gearing arrangements, depicted in a set of woodcuts dedicated to Maximilian I (1526). Source: Hans Burgkmair, *The triumph of Maximilian I* (New York, 1964), p. 93.

creations were the first of such scope and inventiveness to be found in the entire world. His technical achievements, which are often misrepresented, deserve praise for their true worth – not as a set of blueprints for new machines, or as accurate prophecies of the shape of future technology, but as wonderfully imaginative and original explorations of the potential inherent in the enterprise.

Da Vinci may have been unique in the magnitude of his genius but not in his predilection for visionary technological schemes. Schemes continued to proliferate throughout the centuries as technology developed in complexity and influence and as new power sources were developed. There is no evidence that the vigor and popularity of technological visions has diminished. Despite the failure of technology to bring the utopian society promised by its eighteenth- and nineteenth-century promoters, and despite serious problems closely identified with twentieth-century technology, from environmental pollution to nuclear warfare, the visions continue to be produced and to fascinate the public.

The popular press is constantly filled with the promises made by engineers, scientists, and technicians that computers, robots, spacecraft, or whatever will make possible enormous technological advances that will far transcend the layman's expectations. Although such claims are often used for the purpose of self-promotion and self-aggrandizement, they also reflect the great pleasure taken in playing with technological possibility for its own sake.

Impossible Machines

There is always the chance that a future technological breakthrough will facilitate the transformation of the wilder technological dreams into reality. However, the existence and operation of impossible machines can never be altered by future developments in technology because they violate fundamental scientific laws.

Perpetual-motion devices are probably the best known of the impossible machines. For more than fifteen hundred years, mechanicians have offered plans for, and actually built, machines that, given proper construction, materials, and lubrication were supposed to operate forever. Such devices were often expected to do useful work, and to generate more energy than was required merely to keep them running.

The classic version of a perpetual-motion device is a wheel that spins continuously upon its axle without the aid of an external power source. A self-moving wheel is described in the ancient Sanskrit treatise *Siddhanta Ciromani* (A.D. 400–50) and a thirteenth-century illustration of one appears in the sketchbook of Villard d'Honnecourt. In Villard's device an uneven number of mallets hang loosely from the rim of a vertically mounted wheel. They are spaced so that the wheel is constantly unbalanced and therefore always moving.

The Renaissance, which first witnessed so many other manifestations of technological fantasy, was a popular time for the invention of perpetual-motion devices. Often quite elaborate in conception, these might utilize water, air, or the force of gravity, and all were designed as closed cycle operations; for example, the energy generated by a continuous stream of water flowing over a waterwheel would be used to power the pump that lifted the water up to the waterwheel, and so on endlessly. Along with unceasing motion, some inventors also promised the production of excess energy that could be used to run the machinery in a flour mill or serve some other useful purpose. The promise of the boon to humankind of unlimited free power combined with the tremendous challenge of getting the device to work in the first place made perpetual motion an exciting venture for many technologists (Figure III.4).

Interest in perpetual motion grew during the eighteenth century and reached its peak in the nineteenth century as many new machines, along with the newly discovered forces of electricity and magnetism, received widespread attention and as the critical role of steam power in industry and transportation became evident. In England well over five hundred patents were issued for perpetual-motion devices between 1855 and 1903; a similar craze swept through America during those years. Industrialization had brought a new rationale for the goals of the perpetual motionists: Their machines would free nations from the need for scarce natural resources like coal and oil.

It is ironic that precisely at the time when many inventors were convinced they were on the verge of bringing unlimited energy to society, physicists were formulating the laws of energy conservation. Had the perpetual motionists understood these laws, they would have known that it was impossible for any device to produce an excess of energy output over input. But the fact that the first and second laws of thermodynamics implied the impossibility of perpetual-motion machines did not stop inventors from pursuing their dreams. Finally, in 1911 the U.S. Patent Office declared that henceforth all patent applications for perpetual motion machines must be accompanied by working models. Yet the long and futile search for perpetual motion continues to this day. For the enthusiast there is always the hope that some crucial mechanism or circuit can be devised and a workable impossible machine built. That this hope is contrary to the laws of physics and the experience of technology has not discouraged inventors who have long looked upon perpetual motion as the ultimate challenge to their capabilities.

Popular Fantasies

Fanciful machines generated by the literary or popular imagination do not originate in the minds of inventors and engineers and, therefore, intimate that the drive to envision a wide range of possibilities

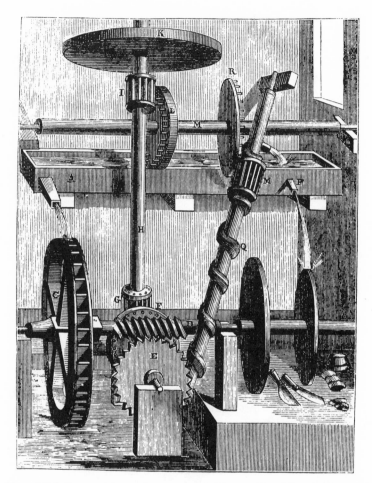

Figure III.4. Seventeenth-century perpetual-motion machine. Water in the trough at the top is discharged over the large waterwheel causing it to turn the Archimedean screw water-pump (Q) that lifts the water perpetually back to the top. The waterwheel also powers two grindstones at the far right on which cutlery can be sharpened. Source: Henry Dircks, *Perpetuum mobile*, 2nd ser. (London, 1861), fig. 151, p. 40.

is not confined exclusively to the denizens of the technical community.

Popular technological fantasies can be traced back at least to the thirteenth century when philosopher Roger Bacon prophesied that large ships, without oars or sails, would navigate rivers and seas;

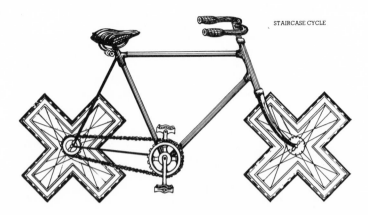

STAIRCASE CYCLE

Figure III.5. A bicycle adapted for stairs. One of the many ordinary objects that have been "improved" for the modern consumer by Jacques Carelman and illustrated in his *Catalogue of unfindable objects*. The catalog contains "improvements" to various items, including plumbing fixtures, furniture, household goods, and sports equipment. Source: Jacques Carelman, *A catalogue of unfindable objects* (London, 1984), p. 56.

vehicles, without animals to pull them, would move rapidly over land; flying machines, with wings that beat like a bird's, would glide through the air; and humans, using diving bells, would explore the bottom of the ocean. Similar prophecies have long enjoyed popularity in the Western world. Industrialization in the nineteenth and twentieth centuries nurtured the predilection for fanciful technological predictions and institutionalized it in the popular arts. Of these arts, science fiction became the single most important source of fantastic machines. Examples include Jules Verne's submarines and spacecraft, H. G. Wells's time machine, Karel Capek's robots, and the starships and laser weaponry that dazzle the modern science fiction moviegoer.

The second element, cartoons featuring fantastic machines, appears to be a minor topic when compared with science fiction but deserves attention because of the cartoons' dual purpose. Rube Goldberg (United States), W. Heath Robinson (England), and Jacques Carelman (France) are three twentieth-century cartoonists who have expanded the repertory of technological fantasy with comic drawings (Figure III.5). Yet a close study of their works discloses that comedy and fantasy are used to conceal the barb of social criticism. Hidden in their drawings are serious statements about the absurdity of an industrial civilization that creates complex machines to accomplish trivial ends and that naively believes all human problems can be resolved by technology. The fantasization of technology is so pervasive in our culture that it can be used to satirize the technological exuberance that brought it into being.

Scientific/technical journalism, as exemplified by *Popular Science, Science and Mechanics, Mechanix Illustrated*, and *Popular Mechanics*, completes the genre. Begun at the turn of the century and aimed at working-class men and boys, these magazines present a strange mix of home improvement tips, plans for workshop projects, technological visions, and the promise of utopia through technology. In the past decade the utopian visions were repackaged in slick and more expensive popular science journals such as *Omni*, which combined science fact and fiction and catered to readers with a higher level of sophistication and education. Yet, no matter what the intended audience, popular science journalism has continued to be one of the purveyors of technological fantasies to a wide public.

The foregoing survey of fantasy, play, and technology yields four general conclusions that contribute to a wider understanding of technology and change: First, the technological imagination is very rich. Hardly constrained by biological or economic necessity, it often exceeds the boundaries of rationality as it contemplates the improbable and the impossible. The fertile technological imaginations create a superfluity of novel artifacts from which society makes selections.

Second, widespread fantasization of technology is primarily a Western phenomenon. The examples cited here are not the result of a deliberate and parochial concentration on European and American sources. It would have been impossible to assemble a comparable set from the records of any other of the great civilizations. The reasons for this Western hegemony are by no means clear, but the process and the results are readily documented. Perhaps the occurrence of technological fantasies at all levels of Western societies can be attributed to certain values that gained ascendancy during the Renaissance: secularism, the idea of progress, and the domination of nature. More will be said about these in chapter IV.

Third, the fantasization of technology calls for a reappraisal of the social role, professional stance, education, and personality of the technologist. Instead of an unimaginative servant duly responding to society's call for necessities, we find a visionary who is apt to offer far, far more than society needs or often wants. If society clings to the former image when the latter is closer to the mark, serious misunderstandings may occur. To take a recent example, if we believe the engineering supporters of nuclear power are giving us a realistic, objective assessment of its costs, benefits, and drawbacks when in fact they are indulging their enthusiasm for a technically attractive form of energy, we are likely to encounter serious problems when we follow their advice.

Fourth, the fantasization of technology is a two-edged sword. Although it contributes to artifactual diversity, it also promotes the unthinking acceptance of technological change as good in and of itself, thus perpetuating the erroneous notion that the solution to most social problems can be found in a set of new technologies.

The understanding that fantasy is a significant element in inventive activity allows us to shift our focus to one of the more traditional sources of novelty in technology — knowledge. Knowledge may take the form of an artifact, or the representation or idea of an artifact, which is transferred from one region or culture to another, or it may be embodied in scientific advances that expand possibilities for making new kinds of things.

Knowledge: Technology Transfer

No society is so isolated or self-sufficient that it has never borrowed at least some aspects of its technology from an outside source. Because humans engaged in normal communications are bound to exchange information about novel techniques or artifacts, general cultural contacts are the oldest means of transferring knowledge about technology from one culture to another. These contacts may be the result of exploration, travel, trade, war, or migration. All of these ensure that the parties concerned will be exposed to new technological opportunities. What is traditional practice for one culture may be an important innovation in a different setting.

In some cases we can date with precision and identify the persons responsible for the introduction of a novelty. On 25 August 1543, three Portuguese adventurers became the first Europeans to visit Japan. They brought with them two matchlocks, muzzle-loading hand guns first made in early sixteenth-century Europe but unknown in Japan. The Japanese were so impressed by these primitive firearms that they purchased them on the spot and set their swordsmiths to work duplicating them. Within a decade, gunsmiths all over Japan were turning out firearms in quantity. The warring feudal factions in Japan, anxious to obtain weapons superior to their swords and spears, encouraged these developments. By 1560 Japanese matchlocks were used routinely in the field, and in 1575 they proved decisive in one of the great military engagements in Japanese history (the battle of Nagashino). The Japanese may have been latecomers to the use of firearms, but they pioneered large-scale manufacture and quickly incorporated guns into their military strategies.

Other technological transfers cannot be dated so accurately, or the agents of their diffusion identified so precisely. Windmills were a significant addition to existing power sources in medieval Europe, yet their origins remain obscure. Vertical-axle windmills may have been used in Persia from about the seventh century A.D.; however, the first European windmills (twelfth century A.D.) are of the horizontal-axle variety. Did diffusion include the tilting of the axle from the vertical to the horizontal plane? Or, more likely, were the horizontal-axle mills an independent European invention? Could the European windmills have been patterned after the widely used horizontal-axle water wheels? No ready answers are available.

Imperialism

Among the specific kinds of cultural contacts that lead to the diffusion of technology, imperialism and colonial conquest are of great importance. Under these conditions the recipient culture has little choice but to accept the technology being offered by its imperial masters. This need not always be deleterious. India under British rule provides one of the best illustrations of how an imperial power, if it is able and willing to do so, can choose to bring the latest inventions to its colonies.

During the two hundred years they governed India (1740–1947), the British introduced virtually every aspect of their material culture to the subcontinent. The majority of the artifacts were brought for British civilian and military personnel and their families, but three of these were key innovations that had a wide and lasting impact upon Indian life: steamboats, railroads, and electric telegraphy. None of these would have reached India when they did, and with the intensity they had, if they had been transmitted by other means.

The age of maritime steam propulsion began in England in 1801 as the steam tugboat *Charlotte Dundas* successfully navigated the Forth and Clyde canal. Eighteen years later the first steamboat traversed Indian waters, a small pleasure craft built by the British for a local prince. Soon experiments with a steam tugboat began in Calcutta harbor, and in 1824 river steamers played a decisive role in defeating the Burmese during the Anglo-Burma war. While some Anglo-Indians looked forward to the advent of steam ocean-going ships that would shorten travel time to Britain, others envisioned a fleet of vessels steaming on the Ganges, carrying passengers and goods along the backbone of the country. Ganges steamers were built, but the problems encountered in navigating the river proved formidable. Freight and passenger rates were never low enough to attract sufficient customers and the service was eventually curtailed; in the meantime, a few thousand Indians were employed on the riverboats, although at inferior jobs. But despite these problems, millions of Indians watched in awe as the steamboats moved upstream without the aid of oars or sails. As a symbol of superior British technology the steamboat was soon to be supplanted by the railroad, an even more potent proof of the ingenuity and technical resources of the British Raj.

The construction of the Indian railway system is one of the largest technological projects ever undertaken by a colonial power. It involved huge investments of time, talent, and money – ninety-five million pounds sterling between 1845 and 1875 alone – and resulted in the fourth largest rail system in the world. Twenty-five hundred miles of track were laid during the first phase of construction ending in 1863. Work continued until the system totaled forty-three thousand miles in 1936. Unlike the steamboat, the railroad did fulfill the goals of providing the Indian people with rapid and low-cost transportation. Even today it remains a crucial link in the travel and transportation system.

A large rail network cannot operate safely and efficiently without an accompanying telegraphic system. Hence, the stringing of the first telegraphic line, spanning the eight hundred miles from Calcutta to Agra (1854), coincided with the construction of the railroad. Work on an Indian telegraphy system had begun just ten years after Samuel F. B. Morse opened his Washington-to-Baltimore line (1844). By 1857 forty-five hundred miles of telegraph lines had been strung across India, and eight years later a series of land and submarine cables connected the subcontinent to the British Isles. Although initially conceived as an adjunct to rail travel, telegraphy soon proved to be even more useful for relaying information.

Altruism did not prompt the British to introduce steamboats, railroads, and telegraphy into India so soon after they first became available in Europe and America. Steam transport over land and water facilitated the movement of British troops and sped the transfer of raw materials to British factories and of British manufactured articles to Indian markets. Telegraphy helped to consolidate the rule of the Anglo-Indian government and to keep it in close touch with London. Yet after acknowledging the truth of these assertions, we must recognize that the transfer of Western technology did more than serve imperialistic ends.

Some of the British who were instrumental in bringing the new technologies to India predicted that peace, goodwill, and plenty surely would be the lot of a people who were exposed to the civilizing effects of modern machinery. Karl Marx, on the other hand, forecast a different future. Once the railroads were built, Marx predicted they would be used by the Indians to exploit their natural resources (coal and iron), to become a modern nation, and to gain the strength needed to overthrow their British oppressors.

Neither vision was correct; Indian life and the Indian economy were not radically transformed by these innovations. What did happen was that the people of India were introduced to Western technology before other Asian societies. For example, India's railway was in operation twenty-five years before Japan's and thirty years before China's. And India opened its first cotton mill in 1851, fifteen years before the Japanese did likewise.

Although the Indians were not placed in command of these new technologies, they were permitted to do the minor jobs that at least brought them within the outer orbit of Western technology. This experience whetted their appetites for modern machines and introduced them to the Western ideas of novelty, change, and progress. When the Indians finally freed themselves from British rule in 1947, they made every effort to join the ranks of industrial nations.

Migration

A second mode of diffusion relies on transmission of skills and artifacts by migrating people. The classic illustration of this is the forced migration of some two-hundred thousand Huguenots (French Protestants) after Louis XIV revoked the Edict of Nantes in 1685 and ended nearly a century of limited religious toleration. The Huguenots, many of whom were highly skilled workers in a wide variety of crafts and trades, took their talents and technical knowledge to England, Ireland, Holland, Germany, and Switzerland. In these countries their innovations contributed to changes in the textile industry, especially for the production of silks, velvets, and laces, as well as to changes in apparel, specifically, hats, stockings, gloves, and ribbons; and they improved the manufacture of fine papers and of blown and cast plate glass.

A migration need not be as large, dramatic, and far-reaching as it was in the case of the Huguenots. The agents of diffusion may be but a small number of knowledgeable persons. This was especially true before the mid-nineteenth century, because published technical drawings and texts were not then generally available. Given these conditions the best way for an individual to learn about a new machine was to deal directly with those who built and operated it.

In 1748 no steam engines operated in America, although they were becoming common in England. When Colonel John Schuyler of New Jersey wanted to obtain a Newcomen engine to drain water from his copper mines he contacted the Hornblowers, a well-known English family of engineers experienced in building steam engines. The engine parts were gathered together in England and sent to America along with young Josiah Hornblower who supervised their assembly at the mines between 1753 and 1755. This English engine inaugurated the age of steam in America.

The Americans were not unique in requiring the assistance of English technical help in erecting and operating their first steam engine. Both Newcomen atmospheric steam engines and Boulton and Watt steam engines were introduced into other lands by experienced engine erectors routinely sent out from England to Germany, France, Holland, Spain, Austria, Sweden, Belgium, Switzerland, Hungary, Italy, Denmark, Portugal, and Russia. In some of the more remote and less industrialized countries, these men found the local populace lacking in the mechanical skills that were taken for granted at home. In 1805 a Boulton and Watt employee in Russia warned that the engine he had just completed there was likely to be harmed through Russian incompetence. The Italians, one reported in 1789, are an "ignorant sett of piple as ever I saw – they kno nothing of mushiniry"[2] During the next few decades Europeans came to know a great deal more about "mushiniry," a direct result of the English mechanics who appeared in their midst to erect steam engines during the eighteenth and early nineteenth centuries.

At about the same time English mechanics were erecting Boulton and Watt engines across Europe, another group of Englishmen, acting clandestinely for the most part, were transferring British textile technology to America. By the late eighteenth century, the British government was aware of the contributions that the technological innovations in the textile industry had made to national prosperity. In 1781 a law was enacted that expressly prohibited the export of any "Machine, Engine, Tool, Press, Paper, Utensil, or Implement" as well as any "Model or Plan . . . Part or Parts thereof"[3] used in textile manufacturing. Meanwhile, across the Atlantic the Americans, having broken their political ties with the mother country, were bent on achieving economic independence by creating a textile industry, and they were not averse to utilizing the forbidden British technology.

The Americans had an ample supply of wood and of metal ores, as well as the skilled craftsmen and mechanics needed to transform those resources into machine components. What they lacked were designers of textile machinery and workmen who knew how to adjust, control, and maintain those machines so that yarn and cloth of an acceptable quality could be produced in quantity from native wool and cotton. Having the actual machines on hand did not suffice if there was no one experienced in assembling and using them. This the Americans learned in 1783 after several key textile machines were smuggled into Philadelphia from England in a disassembled state. After four frustrating years, during which no one competent could be found to assemble them, they were shipped back to England.

David J. Jeremy, who has studied the transmission of British textile machinery to the United States, notes that early attempts at transfer were hampered because before 1812 textile machines were not described in written texts nor were they even the subject of illustrations. Two more decades passed before full information about textile machines was available on the printed page. Under these circumstances the only recourse for Americans was to entice British artisans to immigrate, bringing with them machines, parts, plans, or models, if possible. In some instances a memorized plan sufficed if the individual had good working knowledge of the machine. Successful transfer of textile technology was not achieved until experienced British emigrant artisans were able to put their nonverbal knowledge to use and produce the machines for the American manufacturers.

Practical Knowledge

Because *all* of technology can never be translated into words, pictures, or mathematical equations, the practitioner with a hands-on knowledge, be it of eighteenth-century textile machinery or twentieth-century computers, will always have a role to play in the dissemination of technical innovations. Although much of modern technology can be gleaned from the pages of books, articles, monographs, and patents, the artifacts must be studied at first hand, oral information gathered from persons conversant with the new technology, and the innovations adapted to the recipient economy and culture.

The extent to which technology transfer depends on information that transcends the printed page can be gained from the eighteenth-century account of the introduction of Italian silk-throwing machines into England and from the twentieth-century story of the transistor's journey from America to Japan. The first of these is an example of industrial espionage, the second of the legitimate purchase of a patent license.

Silk production in England dates back to the arrival of emigrant French Huguenot silk weavers in the seventeenth century. The weaving of silk presented no technical problems for the French-born founders of the new industry; however, finding properly thrown silk thread did. Raw silk thrown by hand in England yielded an inferior thread, and the finer silk thread imported from Italy was very expensive. If the infant English silk industry was to flourish, it had to gain access to the water-powered throwing machines used by the Italians to produce superior silk thread cheaply. The Italians regarded the workings of these machines a state secret, and according to the laws of the Kingdom of Sardinia the disclosure of the operation of silk mills was punishable by death. Despite the fact that the Italians jealously guarded the secret of their silk-throwing machines, they permitted the publication of detailed engravings of one of them (Figure III.6) in Vittorio Zonca's *Teatro nuovo di machine et edificii* (*New Theatre of Machines and Buildings*) (1607). This volume, reissued in second (1621) and third (1656) editions, found its way to English readers in the Bodleian Library at Oxford University. Even so, the proprietors of English silk mills did not copy the machine so openly displayed by Zonca. One reason for this was that all of the relevant information needed to build an intricate machine could not and, indeed, still cannot be conveyed in pictorial form. This holds true for seventeenth-century engravings as well as for the best modern engineering drawings. The complete interpretation of a depicted machine can only be made by persons who possess an intimate, practical knowledge of the construction and operation of the actual machine. Therefore, Zonca's illustration posed no threat to the secrets held by Italian silk-thread producers.

After an abortive attempt by Englishman Thomas Crochett to mechanize silk throwing in 1702, a London textile merchant decided that information about the machine must be stolen from the Italians. John Lombe, who came from a family of English textile weavers and merchants and who had a good head for mechanics, was sent to Italy in 1715 to do the job. During a two-year stay, Lombe "found means to see this engine so often that he made himself master of the whole invention and of all the different parts and motions."[4] On his return John's half-brother, Sir Thomas Lombe, constructed a large mill for the mechanical throwing of silk and used John's knowledge of the machinery and techniques. The transfer of this

Figure III.6. Vittorio Zonca's engraving of Italian water-powered, silk-throwing machine. The accompanying caption confirmed some of the more obvious operations, but not the inner workings. Even after the publication of this engraving the Italians continued to monopolize the silk-throwing industry for more than a century. Source: Reprinted with permission of Macmillan Publishing Company from *A theatre of machines* by Alex G. Keller, p. 38. Copyright © 1964 by Alexander G. Keller.

key element of silk-processing technology could not then have been accomplished without the help of an industrial spy who carefully observed the machine over an extended period of time and thoroughly acquainted himself with every aspect of its operation.

Industrial espionage is by no means confined to earlier times when industry was less rationally organized and science was not yet firmly established as a source of technological innovation. In the modern chemical and electronics industries, employees are routinely required to sign restrictive covenants that limit the kinds of technological and business activities they can engage in after leaving their present jobs. Apart from any trade secrets these employees may learn, they are in a position to take with them the special skills, the know-how gained in working with a particular technology over the years.

A second example of technology diffusion and the limitations of the printed page is the transistor. An invention of American and European scientists and technologists in 1947, the transistor was spectacularly exploited for commercial purposes by the Japanese.

Shortly after World War II several young Japanese engineers banded together to form Tokyo Telecommunications Engineering Company, a small electrical products firm. They began by producing an electric rice cooker and a vacuum tube voltmeter, the latter attracting far more buyers than the former. In searching for other marketable electrical appliances the company finally created (1949–51) a magnetic-tape recorder for use in Japanese schools.

While on a trip to the United States in 1953 Masaru Ibuka, one of the founders of Tokyo Telecommunications, heard about the transistor from a friend living in New York City and was informed that Western Electric was about to release the transistor's patent rights for sale. Ibuka knew very little about transistors but decided that it might be the invention his company needed to expand its consumer product lines and keep its technical staff busy.

When Tokyo Telecommunications bought the transistor license in 1954, none of the older and much larger Japanese electronic firms showed the slightest interest. Ibuka sent technicians to America to collect all available technical publications dealing with semiconductors. They visited laboratories to observe transistors being made and talked to the scientists, engineers, and technicians working on all aspects of transistor production. The Japanese team assimilated every bit of written and oral information pertaining to semiconductor technology and then decided to manufacture their own transistors and use them to make a pocket-sized radio receiver. About the time their miniature radio was ready to be marketed in 1955, they changed their corporate name to Sony, which was shorter and smoother sounding than Tokyo Telecommunications.

The Sony radio was not the first small transistor radio in the world – the American-made Regency holds that honor – but Sony showed the electrical giants of all countries what could be done with the transistor. Having captured the lead in commercially exploiting semiconductor technology, Sony and other Japanese electronic firms moved from success to success as they led the world in the production and invention of consumer electronic products.

While the Japanese were making the transistor popular and profitable, American scientists and engineers were carrying out the research that led to the development of new kinds of transistors. The American semiconductor industry supplied the high-technology markets opened by the growth of the computer and the needs of space and military programs. In contrast, the American electronic consumer product industry, reluctant to manufacture transistor devices that would compete with their vacuum tube models, only slowly adopted the new technology. Thus, at a crucial stage in the history of semiconductors, the Japanese were free to enter, define, and dominate the consumer market.

War-torn Japan was an unlikely place for the commercial development of an invention that was the result of the concentrated efforts of some of the best minds of Western science and technology. Japanese scientists were remote, geographically and intellectually, from the latest work in solid-state physics that laid the foundations for the transistor. The engineer–entrepreneurs of Tokyo Telecommunications, however, saw an opportunity in semiconductors that was not perceived by others. Yet before they could seize the initiative they had to come to understand the transistor. Their understanding was gained not only by reviewing printed technical information but also by observing American transistor manufacture and questioning specialists who were immersed in semiconductor technology. In sum, these engineers could not have launched the transistor industry in Japan had they stayed at home and relied solely upon the printed page for their knowledge of transistor technology.

The transistor case study is of interest to students of technology transfer for reasons that go beyond those already mentioned. Semiconductors serve as an example, and by no means an isolated one, of an innovation that was enhanced and utilized by the borrower in ways that were not fully imagined by the initiator. They also prove that the development and commercialization of a product of modern scientific technology can be successfully undertaken by a society whose scientific base is more restricted than that of the one that created the product. In the late 1940s, the Japanese could not have invented the transistor, yet obviously they were adequately prepared to make the best commercial use of it.

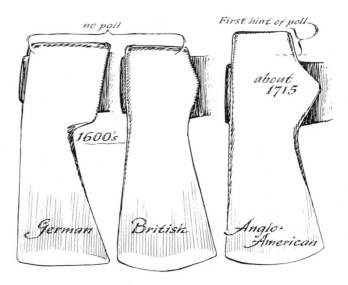

Figure III.7. Axes used in colonial America. The poll on the Anglo–American axe set it apart from European counterparts and provided additional balance and weight that were necessary when cutting clearings in the heavily forested New World. Source: Eric Sloane, *A museum of early American tools* (New York, 1964), p. 11.

Environmental Influences

One final aspect of technological diffusion remains to be explored – the ways in which the natural environment can induce changes in a transferred artifact. A tool or contrivance that has been designed to function in one natural setting often must be altered if it is to work properly in a new environment. Three well-known American artifacts – the axe, steamboat, and locomotive – illustrate the close relationship that may exist between changes in the physical environment and variations in the design of made things.

The first settlers in the American colonies brought with them European-style axes that had served them well in the Old World (Figure III.7). These tools were suitable for hewing, or shaping, logs but not for felling the huge stands of virgin timber in America. The European axe was a light tool without a poll – the extra portion of metal on the side opposite from the bit, or cutting, edge that gave the axe weight and balance.

As the settlers moved westward and southward clearing forests for agriculture and living space, they developed a distinctive American axe for felling large trees. The American axe had a heavy poll that gave the instrument a better balance and provided more weight for chopping thick tree trunks. Rudimentary polls first appeared in the early 1700s; by the 1780s the tool had evolved into a full-fledged American felling axe. These axes, originally made by local blacksmiths, were mass-produced in factories in the nineteenth century.

The American axe was not manufactured in a single, standard pattern or design but was available in a number of models, each one adapted to the particular forest environment in which it was intended to be used. In 1863 one manufacturer listed the following varieties of felling axes: Kentucky, Ohio, Yankee, Maine, Michigan, Jersey, Georgia, North Carolina, Turpentine, Spanish, Double-bitted, Fire Engine, and Boy's-handled. Thirty-five years later the list exceeded a hundred.

The axe was carried from the Old World to the New where it underwent a transformation. The American steamboat, on the other hand, was transformed in the short move it made from the east coast to the rivers of the Mississippi basin. The steamboat, first used on the rivers along the Atlantic seaboard near the end of the eighteenth century, reflected its geographical origins in two respects. First, its form was derived from that of sea-going ships; second, it was originally designed to travel the waters of the Hudson River and Long Island Sound. The decision to put it to work on midwestern rivers required the steamboat manufacturers to make a series of fundamental design changes.

The hulls of sailing ships were deep, well rounded, and fitted with a projecting keel on their bottoms. They were heavily planked and framed with large timbers to withstand the stress of storms at sea. Cargo holds and cabins for passengers and crew were situated below the main deck and acted as ballast to offset the top-heavy masts, sails, and rigging. The low center of gravity and sturdy construction enabled the sailing ships to withstand the rough weather and heavy seas. The builders of the first steamboats borrowed key structural elements from wind-powered vessels.

Structural features that were a necessity for marine craft were either useless or counterproductive on boats plying western rivers. These streams were relatively shallow and seldom, if ever, had large damaging waves. In case of a storm a steamboat was never far from shore. Sails were not needed on a steam-powered vessel nor could they have been used effectively within the narrow confines of a river.

Because of the different conditions found on inland waters, the sea-going model of the eastern steamboat was transformed into the river steamboat in less than fifty years. By the 1850s the classic steam-powered river boat emerged. It was a relatively light craft without the planking and additional structural members called for in sailing ships and incorporated in the earliest steamboats. The hull, radically different because the keel was unnecessary, was made broad and flat-bottomed and its overall length was increased. The additional length and breadth of the boat meant that the area of the hull resting on the surface of the water was larger, and consequently the vessel was more buoyant. Hence, the boats drew less water and could navigate shallow rivers at faster speeds while carrying a full load of cargo and passengers. The decreased depth of the hull required that power plant, passengers, and cargo be placed above the main deck. All of these were housed in a superstructure that gave river steamboats their characteristic boxy appearance.

Steamboats in the Mississippi Valley reached their peak in the 1850s only to be challenged, and defeated, by the railroad. The first steam locomotive to operate on American tracks was imported from England in 1829. Not long afterwards Americans redesigned the English locomotive to adapt it to meet peculiar American needs. The divergent designs of English and American locomotives cannot be explained solely in terms of the physical environments of the two countries; however, one feature of the American locomotive, the leading truck, was the direct result of different environmental conditions.

The first English locomotives used large driving wheels connected directly to the steam power plant and attached to a rigid frame beneath the boiler. The driving wheels, which moved the locomotive and bore the brunt of its weight, operated best on straight sections of track or those with wide curves. Because they were unable to pivot they were ineffective on sharp curves. However, relatively straight and level roads were common in England because of the nature of the terrain and the willingness of English railway builders to invest money in tunnels, cuts, and bridges so that the rails could run directly through or over an obstruction instead of detouring around it. The English locomotive was well matched to the geometry of its rail system.

In America, where the terrain was more varied, a different theory of railroad layout prevailed. In what became known as the "American method" of railroad construction, railways were built cheaply and quickly with sharp curves, steep grades, and poor roadbeds. Tunnels were avoided, and bridges were built only as a last resort and then of wood.

The English locomotive did not function well on American roadbeds. Early in the 1830s, John B. Jervis of New York State proposed the addition of four small independently mounted wheels at the front of the locomotive to help carry the locomotive's weight and guide it along the track. These wheels were not connected to the engine's driving rod but were attached to a truck that pivoted freely as the locomotive rounded a sharp curve.

Jervis's proposal led to the first fundamental revision in locomotive design made in America. It enabled locomotives to traverse more easily the serpentine tracks that were being laid across the American landscape.

This discussion of the diffusion of the axes, steamboat, and locomotive has stressed artifactual modifications generated by the natural setting: forests, rivers, and terrain. Little has been said here about the larger social, political, economic, and cultural environment and its influence on these artifacts. A concern for artifact modification within a more broadly conceived notion of the environment has been raised by historians James E. Brittain and Thomas P. Hughes. They studied the variations induced in the electric power generator as it spread throughout America and western Europe between 1870 and 1920. The design, manufacture, and distribution of absolutely identical generators proved to be impossible because each country had a slightly different set of requirements. The machines were consequently altered to meet nationally expressed wants and desires with the result that a number of variant generators were built, each model especially suited to its uses in a specific country.

A brief look at the artifactual world today reveals that automobiles, telephones, household appliances, and television sets, to name a few, have undergone changes similar to those of the generator. Each of these artifacts was altered to conform to changing circumstances and patterns of usage as they were introduced into different countries. The automobile, to take an obvious example, was adapted to national driving customs, road conditions, fuel costs, safety regulations, and terrains. A comprehensive theory useful in explaining artifactual variation through adaptation is not yet available.

Knowledge: Science

With the exception of the transistor, the artifacts already discussed in this chapter were largely untouched by the advancement of science. In the twentieth century, however, science has come to play a much larger role in the creation of technological innovations and hence deserves separate treatment. Proponents of scientific research have exaggerated the importance of science by claiming it to be the root of virtually all major technological changes. A more realistic and historically accurate assessment of the influence of science on technological change is that it is one of several, interacting sources of novelty.

The invention of the atmospheric steam engine and of radio communications are two distinct technological events that warrant detailed study because of what they reveal about the nature of the interaction of science and technology. Several general conclusions drawn from that study can be applied to other examples of inventive activity in which science plays a part. First, the connection between science and technology is complex and never simply hierarchical. Second, the scientific knowledge that spurs technical innovation need not be the latest nor need it appear in its purest form; second- or third-hand conceptions of scientific advances can and do serve technology well. Third, science dictates the limits of physical possibilities of an artifact, but it does not prescribe the final form of the artifact; Ohm's law did not dictate the shape and details of Edison's lighting system nor did Maxwell's equations determine the precise form taken by circuitry in a modern radio receiver.

A search for the origins of the steam engine raises the question of whether Thomas Newcomen could have invented his atmospheric engine without the help of science. In the prehistory of the steam engine technological elements, not scientific theories, predominate, except for the idea of the vacuum. The study of the vacuum was not part of the crafts tradition associated with the making of machinery but grew out of the concerns of early scientists who were investigating the physics and metaphysics of spaces devoid of matter.

Aristotle's claim that a vacuum could not exist in nature was challenged in the seventeenth century by Galileo Galilei, Evangelista Torricelli, Blaise Pascal, Otto von Guericke, and others who contributed to the development of pneumatics. They proved that the earth's atmosphere exerts a pressure, built pumps capable of evacuating the air from small containers, and studied the vacuums they produced in their laboratories. Whereas some sought to learn whether a vacuum would support life or transmit light or sound, others considered its possible utilitarian applications. One of the latter was Denis Papin (1647–ca. 1712), a French scientist who conducted some of the earliest experiments with steam, evacuated cylinders, and pistons.

Papin's first trials, made on the recommendation of the Dutch scientist Christiaan Huygens, used the explosion from a small charge of gunpowder to expel the air from an upright cylinder fitted with a piston and valves. The exploding gunpowder was not meant to drive the piston but was designed to rid the cylinder of its air so that the weight of the atmosphere pressing on the piston's upper surface would cause it to move downward into the partially evacuated space. The gaseous products left behind by the explosion of gunpowder made it impossible to create anything close to a perfect vacuum in Papin's cylinder. Therefore, he next tried steam in his apparatus.

A small amount of water was poured onto the bottom of the cylinder, and the piston was forced down by hand until it touched the fluid's surface. When Papin applied fire directly to the thin-walled cylinder, the water was heated and converted into steam. The expansive force of the steam moved the piston upward slowly. This weak upward thrust of the piston proved the less important motion. Once the piston had reached the upper limit of its travel, it was held immobile; the flame was removed, the cylinder cooled, and the steam condensed. Thus beneath the piston was a vacuum, and above the piston the weight of the atmosphere. Next the piston was released. It plunged downward with a powerful force that Papin was able to measure (Figure III.8). Papin had discovered the key principle of an atmospheric engine and recognized that, given cylinders and pistons of appropriate size, it would be possible to do useful work with them. In the paper he published describing these experiments, he suggested that the power of the atmosphere be harnessed to lift water and ores from deep mines, propel bullets, and move ships without sails.

Denis Papin received a doctorate in medicine, taught mathematics in a German university, published papers in the leading scientific periodicals, and was in close touch with the finest scientific minds of England, France, Germany, and Italy. Although his experiments with steam had technological implications, they were carried out within a scientific setting, grew out of inquiries made by scientists into the nature of the vacuum, and were reported in a scientific paper. Although Papin was a skilled mechanic who constructed his own apparatus and showed an interest in practical problems such as the processing and preservation of food, he was far removed from the kind of craft and industrial milieu that molded his contemporary Thomas Newcomen (1663–1729).

Newcomen, a man of little formal education, established himself as an ironmonger in Dartmouth, England, in 1685. As an ironmonger he both sold industrial hardware and fashioned iron, brass, tin, copper, and lead items for his customers. His craft, a highly specialized one, was a forerunner of modern mechanical engineering.

Newcomen was successful in a trade that brought him into contact

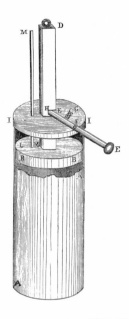

Figure III.8. Denis Papin's steam apparatus, 1690. The piston (BB) is held immobile by a rod (EE) inserted in notch H. The space below the piston is filled
with steam made by heating some water in the cylinder. The next step is to cool
the cylinder and condense the steam, thus creating a partial vacuum. When the
rod (EE) is removed the weight of the atmosphere causes the freed piston to plunge
downward forcefully. Papin realized that this downward stroke could be used to
do useful work. Source: James P. Muirhead, *The life of James Watt* (New York,
1859), p. 107.

with men who built and used machines of all kinds, especially with
those in the Cornish and Devon mining and quarry industry. Therefore, he was well suited to take Papin's vague idea of an atmospheric
engine and turn it into a functioning physical device that could be
profitably used to pump water out of mines.

One of the historian's tasks is to determine exactly how the English
ironmonger learned about the French scientist's steam experiments.
A few scholars have claimed that there was no connection between
the two, that Newcomen and Papin independently invented similar
devices. Simultaneous discoveries are by no means uncommon, but
they are usually made when two or more researchers are working
near the frontiers of their shared specialty. There is nothing in the
historical record to indicate that Newcomen shared the intense sci-

entific interest in pneumatics that led Papin to experiment with gunpowder and steam.

At this point a chronology of events becomes helpful. Newcomen's first successful engine was erected in the Midlands in 1712, and contemporaries claimed that he had spent the past ten years perfecting it. In 1690 Papin published, in Latin, his paper relating the experiments with steam; a French version followed in 1695. There was no early English translation of Papin's paper, but a review of the French one was printed in the March 1697 issue of England's premier scientific journal, *The Philosophical Transactions of the Royal Society of London.*

The English review summarized Papin's paper in a short paragraph that began: "The fourth [paper] shews a Method of draining Mines"[5] and it went on to describe succinctly Papin's steam—atmospheric pressure experiments and emphasized their practical importance. We have no definite proof that Newcomen discovered this review, but it is likely that if he did not, someone else, knowing his interest in mines and machines, called it to his attention. Unlike Papin who was a Fellow of the Royal Society and a contributor to its journal, Newcomen had no formal association with that famous scientific organization; however, he did have acquaintances who were Fellows or in some way connected to the society.

Newcomen had neither the education nor inclination to pursue the disinterested study of the vacuum, and Papin had neither the interest nor the technical knowledge and imagination to transform his small-scale laboratory demonstration into a practical engine. The work of the two men complemented one another nicely: On one side we find a man of science bending toward the utilitarian; on the other side a man of practice learning of a scientific experiment and making from it a machine to pump water from mines.

It would be a mistake to conclude that Papin, in discovering the principle of the atmospheric engine, showed greater originality and genius than did Newcomen, who first used it in a fully operating model. Nor is it correct to assume that Newcomen merely put theory into practice, that he did what was obvious in following the lead of Papin's work. The magnitude of Newcomen's achievement is revealed when the engine he designed is compared with Papin's bare cylinder and piston. It requires a giant step to move from a piece of laboratory apparatus to the large, intricate machine erected by Newcomen. There is very little in Papin's apparatus that could have served as a guide to the English inventor as he contemplated the making of an atmospheric steam engine.

After an inventor chooses the mechanisms and decides how they should interact in the completed machine, the historian is able to reconstruct the paths that led to the invention. However, the inevitable development of the machine cannot be recounted, because the particular configuration of mechanisms we know as Newcomen's engine only became inevitable *after* Newcomen produced it. Before that there was no clear path for him to follow; there was no one right, self-evident, logical, or scientific way to design an atmospheric engine. Using his knowledge of materials and mechanisms, Newcomen devised an excellent and long-lived solution to a very real problem. His engine was successful in pumping water from mines in Great Britain, Europe, and America, and it survived the introduction of the more efficient Watt engine and remained in use until the beginning of the twentieth century.

Although we do not know the exact sequence of events that led Newcomen to his solution, we can identify the changes he made and the novelties he introduced as he incorporated Papin's cylinder and piston into his conception of an engine. To begin, consider the cooling of the hot, steam-filled cylinder: Papin threw cold water onto its exterior; Newcomen injected the water into the cylinder's interior, thereby cooling it more rapidly. Papin heated water in the cylinder; Newcomen added a separate boiler that could be kept constantly hot while the cylinder was alternately heated and cooled. With respect to scale, Papin's cylinder was 2.5 inches in diameter, whereas Newcomen's first cylinder measured 21 inches across; the larger size required special care in the production and fitting of cylinder and piston. Finally, Papin was content to open and shut valves by hand; Newcomen had to devise valve gear that operated automatically so that the engine could complete twelve to fourteen strokes per minute without human intervention.

The piston presented its own problems. Newcomen's walking or pivoting beam, with one end attached to piston rod and the other to the pump, had no counterpart in Papin's experimental apparatus. Steam raised Papin's piston; the weight of the pump forced the beam to lift the piston to the top of the cylinder in the Newcomen engine. Some of the elements of the walking beam can be traced to contemporary machinery, but Newcomen's invention was a synthesis, not a combination, of them.

What was true of the walking beam was also true of Newcomen's completed engine. It was a synthesis that wedded mechanical technology to the new science of pneumatics. The engine that resulted from this novel alliance initiated an immensely important series of heat engines, a series that continues to generate new artifacts nearly three centuries later.

Recent historical scholarship has called attention to a mode of thinking that is prevalent among those working in technology. In an influential essay, Eugene S. Ferguson has argued that visual, non-verbal thought dominates the creative activity of the technologist – a kind of thinking that is done with images. The visualization and assembly of the components of the machine first takes place in the technologist's mind and is refined by numerous sketches and models. Only then is the technologist ready to describe, write about, or construct a device in the real world. The process of nonverbal thinking is central to the work of engineers and technicians but much less so to scientists, who are more likely to manipulate concepts, mathematical expressions, or hypothetical entities. The prehistory of the steam engine includes the conceptual understanding of the vacuum, an intellectual activity familiar to Papin. The creation of the atmospheric engine called for a different kind of knowledge and a different mode of thinking, which was most familiar to Newcomen. Without the scientific study of the vacuum, we would not have had an atmospheric engine. Without the technologist's ability to visualize how machines work, how they might be changed, and how new ones might be designed to do new things there would have been no atmospheric engine.

Another example that illuminates the interaction between science and technology is radio communication. Radio communications ultimately depended on the theory of electromagnetism developed by James Clerk Maxwell (1831–79). During the twenty-five years between 1854 and 1879, this Scottish physicist reformulated in mathematical terms most of what was known in his day about electricity and magnetism, including Michael Faraday's (1839–55) theories positing the existence of magnetic and electric fields. As Maxwell developed mathematically based laws of electromagnetism, he found that his equations needed a new term to maintain consistency. This term, demanded by the mathematics and not based on experimental evidence, he interpreted as a current or wave that flowed through space. Named a "displacement current," it gave rise to a changing magnetic field that in turn created a new electrical field. Hence, there was a sequence of changing magnetic and electrical fields, one succeeding the other, and one inducing the other. Because all of this took place in space, the traveling fields could be thought of as electromagnetic waves propagated through space at the velocity of light.

Up to this point Maxwell had conducted no experiments, although his mathematics did rest upon, and was in agreement with, known electrical and magnetic phenomena. Nor was Maxwell moved to make any effort to verify the existence and determine the velocity of his hypothesized waves. The rigor with which he had pursued his mathematical thinking and the close fit between his equations and the body of knowledge surrounding electricity and magnetism convinced him that he need not do so.

A man who was bold enough to predict the existence of a radically new entity and yet saw no need to confirm its reality was certainly not likely to concern himself with its technological or commercial possibilities. Maxwell, the theoretician, had very little use for the application of physical principles. In 1878 when he first encountered Alexander Graham Bell's new telephone, he disdainfully commented that his "disappointment arising from its humble appearance was only partially relieved on finding that it was really able to talk." To him Bell's instrument consisted of familiar parts that could have been "put together by an amateur."[6]

Maxwell's theory of electromagnetic waves may have been a great intellectual achievement, but it was not an entirely convincing one as far as some English and continental scientists were concerned. Then in 1887, twenty-three years after Maxwell's paper on the subject was first published, German physicist Heinrich Hertz (1857–94) experimentally verified the existence of electromagnetic waves. In order to do so, he had devised a radiator (transmitter) of the waves and a detector (receiver), so that he could prove that they traveled as claimed, but he used simple electrical equipment that could have been found in most well-stocked laboratories of the time, including Maxwell's. Hertz's transmitter was a battery-powered induction, or spark-coil, similar to the ignition coil in a modern automobile, with an adjustable spark gap and with two attached flat, metal plates that acted as a dipole antenna. His detector was a loop of wire broken by a small gap. The oscillating charge at the transmitter's gap created electromagnetic waves that were radiated through space. On reaching the detector they caused stationary electrons in the wire to move and a spark to appear at the gap in the loop.

Spark radiotelegraphy was born in Hertz's laboratory. With slight modifications his apparatus could have been adapted to send coded messages. But Hertz was not interested in communications technology; he was a scientist verifying a crucial part of Maxwell's theoretical work. Popular contemporary accounts of Hertz's experiments mentioned its possible practical uses, but the German scientist made no allusion to that aspect of his research.

At about the same time that Hertz was conducting his experiments with electromagnetic waves, English physicist Sir Oliver

Lodge (1851–1940) was engaged in similar work. Lodge's subsequent research with Hertzian waves is important because it represents the first steps, albeit halting ones, in the direction of the development of wireless telegraphy.

Both Hertz and Lodge built transmitting and receiving apparatus to demonstrate certain scientific principles; however, Lodge was more intrigued by technological problems than was his German counterpart and was willing to try his hand at their solution. His investigation of electrical waves, for example, grew out of research aimed at improving lightning rods that offered inadequate protection during thunderstorms. But despite his practical interests and his superior knowledge of electromagnetic radiation, Lodge was no early convert to the idea of wireless telegraphy.

In 1892 another English physicist, Sir William Crookes, wrote a popular magazine article praising the wonders of the waves recently discovered by Hertz. In the future, he prophesied, they might allow us to control the weather, grow better crops, and light homes without the use of transmission wires; at present they could be used to create a telegraphic system that needed no wires, posts, cables, or costly appliances. Historian Hugh G. J. Aitken believes that 1892 marked a watershed in the development of radio communication. Previously, experimentation with electromagnetic waves was done to validate Maxwell's theory. After 1892, however, the experimenters shifted toward signaling systems, the refinement and invention of apparatus, and commercial developments that called for patent applications and not scientific papers.

In 1894 Lodge demonstrated his transmitting devices at the annual meeting of the British Association for the Advancement of Science. He sent signals in Morse code across a distance of 180 feet and discussed the possibility of radio telegraphy. At that moment Lodge was in full command of the best contemporary scientific and technological knowledge on the subject of wireless transmission. Moreover, he was at work on an aspect of it that was to have an enormous influence in the future – selective tuning. This innovation would confine the senders of radio communications to narrow, assigned frequencies of operation that would limit or eliminate signal interference. In 1897 Lodge somewhat reluctantly applied for patents covering his earlier work, and he even entered into an agreement with a firm to manufacture radio equipment he had designed. Yet, after all of this, he remained the physicist that he truly was. Concerned about the restriction of knowledge through patents, and intrigued by new areas of physics that were opening up at the end of the century, Lodge never became the mover behind a commercially feasible wireless operation. He had more than enough scientific and technological background to do so, but he lacked the ambition, zest for business involvement, and public presence it demanded.

None of the aforementioned qualities was missing in Guglielmo Marconi (1874–1937). The son of a wealthy Italian father and an Irish mother, Marconi had a limited formal education but did study informally with Augusto Righi, a physicist at the University of Bologna who experimented with short wavelength Hertzian radiation. In 1894 the twenty-year-old Marconi, with Righi's help, assembled an apparatus for electromagnetic wave transmission and by 1895 he had achieved a signaling distance of 1.5 miles. These earliest trials are characteristic of Marconi's later efforts. First, his approach was highly empirical. It could hardly have been otherwise because his knowledge of physics was far inferior to that possessed by a Hertz, Lodge, or Righi. Second, he was obsessed with extending the physical range of his apparatus. He focused on sending his signal over greater and greater distances, and he became convinced a commercially viable radio communications system was possible.

In 1896 Marconi moved to England, intending to pursue the commercial exploitation of Hertzian waves. He did so immediately and dramatically by applying for and receiving a patent on a "method of transmitting signals by means of electrical impulses."[7] This patent, the first issued anywhere in the world for radio telegraphy, encompassed virtually the entire technological application of the scientific work of Maxwell and Hertz. Marconi brought little that was new or original to the patent, but he was the first to claim existing methods, equipment, and circuits as *property*. Under British law this claim was all that was needed to justify his right to a broad patent covering electromagnetic signaling. Only after Marconi had seized the initiative did Lodge belatedly seek patent protection for his research (1897).

Antenna design was the only area of the patent in which Marconi could rightly claim to have made original contributions. His early concentration on distance-signaling prompted him to experiment with different antennae, especially those of the ground-plane or grounded-vertical variety. He did not invent this type, although he was the first to incorporate it in a transmitting system. His experiments with antennae were conducted on a trial-and-error basis. Antenna design today is still something of an art, but at the turn of the century it had a minimal scientific foundation, and Marconi was pushing beyond current scientific frontiers as he hurried to his goal of commercial radio communication. He could not afford to wait for the theoretical justification of a particular antenna configuration.

With the help of wealthy British relatives, Marconi founded the Wireless Telegraph and Signal Company in 1897, an action that once again set him apart from the scientists who were doing research on electromagnetic waves. The company was well funded and staffed, but because it was a pioneering business there was some confusion as to precisely what markets it was to serve. This matter was resolved by Marconi the entrepreneur, the man who wanted to define, seek out, and create new markets. Initially Marconi's firm manufactured radio equipment for sale to others who were expected to set up, operate, and maintain their own system. Its first customers were the British army and navy, who could well afford to train staff for this purpose. But by 1900, because it was evident that the large maritime industry could use radio for ship-to-shore communication, the company established a subsidiary to serve this new market. It trained radio operators who worked at Marconi-owned installations on shipboard and at shore stations. Accordingly, radio transmission service, not equipment, became the product offered by the Wireless Telegraph and Signal Company.

The commercial moves created new technological challenges for Marconi and the technical staff he hired to increase the range and effectiveness of the radio signals. In 1900 signaling distance was limited to 150 miles, but within a year Marconi was attempting to send wireless messages across the Atlantic Ocean. He succeeded in doing so by using very large antennae and high power in his spark-gap transmitter. By sending signals over distances in which the curvature of the earth assumed a critical importance, Marconi was once again working at the forefront of science. As the wireless network grew in size, and more transmitters and receivers came into use, the need for selective tuning arose, and Marconi was finally forced to deal with the problem that Lodge had considered in the 1890s.

The early history of the Marconi company is one in which technology went in advance of, and most often without the help of, science. Marconi was not applying scientific knowledge to the solution of technical problems, he was providing technological solutions for problems not yet comprehended by the scientific community. It is both ironic and just that Marconi was awarded the 1909 Nobel prize for physics. He shared it with German physicist Ferdinand Braun who had designed a sparkless antenna circuit that appreciably increased the transmission range. In explaining the basis for the award, the Nobel committee acknowledged the brilliant theoretical work of Faraday, Maxwell, and Hertz but concluded that Marconi alone had demonstrated the "ability to shape the whole thing into a practical, and usable system."[8]

The cases of the atmospheric steam engine and radio communications are insufficient in themselves to yield a set of general rules covering the relationship between science and technology. They should, however, serve as a warning against the easy acceptance of oversimplified accounts that endow the scientist with all of the creativity and leave the technologist to implement the knowledge handed down from above. The intellectual achievements of Newcomen and Marconi were every bit as impressive as those of Papin and Hertz.

The two cases also reveal the important role played by intermediaries in the transmission of information. Papin was close to those scientists doing pneumatic research, yet his wider interests included at least considering the technological applications of that research. Therefore, Papin was more accessible to Newcomen than was a Torricelli. Similar roles in radio were assumed by Hertz, who turned Maxwell's theory into a laboratory demonstration, and Lodge, who took the laboratory demonstration and moved it toward the technology and business of wireless telegraphy. It was left for Marconi to complete the process by incorporating all of past knowledge with his own considerable insights into the science and technology of radio signaling. At each stage of the exchange, there was a two-way flow of information between the scientific and technological communities; each had something of value to learn from the other. As a result, both technology and science were altered by the invention.

The steam engine and radio communications are post-Renaissance inventions. That is to be expected because modern science, which helped to make these innovations possible, was a product of sixteenth- and seventeenth-century European culture. Before the Renaissance, and for several centuries thereafter, technological advances were achieved without the help of scientific knowledge. The situation changed in the late nineteenth century with the foundation of the science-based chemical and electrical industries. However, this does not mean that twentieth-century technological and industrial growth is entirely dependent on scientific research. Key features of the modern material world continue to be shaped primarily by technology.

CHAPTER IV

Novelty (2): Socioeconomic and Cultural Factors

In this chapter, examples from economics, anthropology, and history illustrate a wholly different set of explanations for the emergence of technological novelty. Of these new explanations, the ones based on socioeconomic factors are the best known and the most fully developed. Their popularity and advanced state of development stem from their connection with economic theory and the Marxist interpretation of historical change. Despite the sophisticated theories and empirical findings that can be marshaled in support of socioeconomic explanations, their drawbacks become evident under critical scrutiny. Therefore, in the end we turn to a broad-based interpretation of innovation that stresses cultural attitudes and values.

Making Things by Hand

"All imitation," writes social anthropologist H. G. Barnett, "must entail some discrepancy."[1] No matter how dedicated a copyist is faithfully duplicating an original, the copy always differs from its model. This is true even when the copyist and the original maker are one and the same person; the mindset, materials, tools, and working conditions are all slightly different and that makes exact reproduction impossible. When more people are involved in the copying process, the number of deviations from the original are even greater.

The impossibility of imitation without discrepancy also holds true for mass-produced artifacts. Random variations are admittedly quite small, but they do exist despite the rigid controls employed by modern industry. On close examination, supposedly identical beverage cans can be distinguished by the arbitrary marks left by the manufacturing process: a rougher or smoother finish on the pull tabs; variations in the incised and relief lettering on the can tops; a misshapen head on the rivet that attaches the pull tab to the top; differences of ink thickness in the wraparound lithographed label, especially where the ends overlap; and deviations in label coloring and lettering. Random variations in mass-produced objects underline the point that variability is the absolute rule in the made world.

None of the minor variations found in either handcrafted or machine-made objects is likely to accumulate and lead to an important innovation; however, human intervention can guide the variations toward a new artifact. Something of this sort may have occurred during the long, slow evolution of lithic technology. Random variations that occurred in the making of stone tools may have suggested new forms and functions that were exploited later. This process may account, in part, for the movement from multipurpose to specialized hand tools.

Some variations that arise during the routine fabrication of things are within the control of the maker. Occasionally an individual makes a deliberate attempt to be different, to break the pattern, to innovate. These routine innovations are more easily studied in handicraft than in machine production because innovation and implementation are usually carried out by a single individual or at best by a few persons.

Anthropologists investigating the making of baskets and pottery by skilled men and women have helped to clarify the nature of routine innovation and its reception in traditional societies. The art of basketry, for example, reached a very high degree of excellence among the Yurok-Karok Indians of northwestern California. Lila M. O'Neale, who early in the 1930s observed Yurok-Karok women weaving baskets, found that virtually all of them were sensitive judges of what constituted a good or bad basket. They not only made baskets, they also thought about them as utilitarian and aesthetic objects. Because material was limited by the natural products traditionally used (twigs, roots, and grasses), and form was limited by the utilitarian nature of the containers, designs interwoven to produce decorative effects offered the clearest opportunity for innovation. O'Neale had no difficulty in locating innovators, women who created new styles and themes of ornamentation. They were admired for their inventiveness, and their designs were incorporated into baskets made by co-workers, but everyone understood that this innovation was merely a passing fad. In the long run the novelties would be forgotten, the older patterns would prevail, and Yurok-Karok basketry would continue along traditional lines.

Ruth L. Bunzel's studies of Pueblo potters (1920s) and May N.

Diaz's work on the Mexican potters of the town of Tonalá (1960s) both stress a conservative outlook that dominates the form and decoration. Diaz found that because of the premium placed on copying and imitation, potters resisted the efforts of a local ceramics museum to encourage the invention of new designs. Yet a few Tonaltecan potters were held in high esteem as innovators. Similarly, Bunzel mentions potters Julian and Maria Martinez of San Ildefonso, New Mexico, who introduced new technical processes, shapes, and ornamentation. The couple's innovative styles became so popular, they displaced the older ware and its ornaments.

A socioeconomic theory that predicts the kinds of craftworkers who are likely to innovate has been formulated by sociologist George C. Homans. Homans assumes that the craft is pursued to earn a living and consequently that social and economic factors combine to impede or accelerate an existing tendency to creativity. For the sake of analysis, he divides craftworkers into three status categories: high, middle, and low. Of these groups, the middle rank is the least likely to introduce an innovation. Those at the bottom of the hierarchy have little to lose by making novel artifacts; they can only hope it will help sales and bring attention to themselves. If they fail they cannot slip to a lower ranking. Workers with the highest status innovate in order to prove their superior abilities and maintain their position of leadership. They have the leisure, authority, experience, and freedom to experiment. Caught between these two innovating classes, the members of the middle group take a conservative stance. They have more to lose than those below and do not feel the pressures to perform of their superiors. Fearing that innovation might jeopardize their position, they uphold the status quo in craft practices.

This theory has been corroborated by studies made in Africa and Latin America. Among the Ashanti of what is now Ghana in West Africa, wood-carvers situated in the middle group do not innovate. Rather than waste time and resources on novel ventures, the average carver turns out replicas of well-known traditional pieces for which there is a ready market. Innovation, which is left to the high and the low, takes two radically different forms. The lowest-ranking carvers, with few skills and a marginal position in the market, strike out on new paths by mimicking the styles of other African tribes or producing fantasy objects that cater to a Westerner's notion of African "primitivism" and have no basis in indigenous art. The master carvers at the top of their craft innovate within the confines of Ashanti tribal art. In so doing, they gain respect among connoisseurs as founders of new aesthetic trends.

The studies of Latin Americans illustrate what occurs when the economic risks of innovation are removed. In some settings researchers find pottery workshops in which management pays wages to the potters and assumes all financial risks. Under such conditions traditional potters in Yucatan, Mexico, regularly produce innovative pieces on demand for the souvenir-shop proprietors who employ them. Similarly, Peruvian potters in a sheltered, government-sponsored artisan school were freed of all economic consequences of innovation. They responded by developing a number of new forms and clay-working techniques. Left to the rigors of the marketplace, neither set of potters would have dared to experiment with novelty so boldly, unless they belonged to the ranks of the highest or lowest status groups.

Homans's theory deals with the socioeconomic constraints on innovation but not with the sources of novelty; and it does not cover innovative behavior when market forces are not operative, as in production for immediate domestic consumption or for personal pleasure. In traditional societies craft education and cultural values are likely to work against change no matter what the goal of the producer.

Suppose that traditional potters, basket makers, or wood-carvers had been given a wholly new kind of material to work with. Would they have responded to it by creating innovative artifacts? Probably not. The extent to which innovation is promoted by the introduction of new materials has been greatly exaggerated. Instead, given a change of material, the workers are more likely to expend extra effort to accommodate the new material to the old form. So it was when metal first became available for use in making hand tools. Tool shapes and types that had evolved in stone were transferred to copper and bronze. The new metal tools long bore the impress of their lithic prototypes.

Each new use of metal recapitulated the process just described. The first all-iron bridge ever built was erected across the Severn River at Coalbrookdale, England, in the late 1770s. Although this famous bridge served as a symbol of the new uses for iron, it was modeled on and built according to woodworking practices. Its cast-iron joints were carefully dovetailed as if they had been chiseled from wood, and its sections were fastened together by iron keys and screws rather than rivets or nuts and bolts. Within a decade another iron bridge was constructed near Sunderland, England. In this structure, cast-iron boxes, roughly the size and shape of masonry blocks, were assembled as if they were so many pieces of dressed stone.

The regularity with which new materials are handled and worked

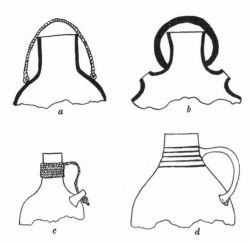

Figure IV. 1. Skeuomorphic structures and designs in Belgian Congo (now Republic of the Congo) pottery. Pots a and c have traditional cord carrying-handles. Pots b and d were made with pottery handles based upon the form and design of the cord prototypes. The pottery handles are skeuomorphs. Source: R. U. Sayce, *Primitive arts and crafts* (Cambridge, 1933), p. 90.

in imitation of displaced, older ones has led archaeologists to coin a word to designate the phenomenon: *skeuomorphism*. A skeuomorph is an element of design or structure that serves little or no purpose in the artifact fashioned from the new material but was essential to the object made from the original material. Architecture offers some of the best known examples of skeuomorphism. Many features of wooden building were repeated in stone by the ancient Greeks when they began masonry construction. Wooden support columns were transformed into stone columns, mortise-and-tenon joints were reproduced in masonry, wood-carved ornamentation was carried over into stone, and the ends of wooden roof joists protruding under the eaves became the familiar Greek ornamental dentil – rows of regularly spaced masonry cubes placed under the eaves of a stone edifice.

Pottery is a craft particularly noted for skeuomorphs (Figure IV. 1). The painted or incised designs encircling clay pots are often the last vestiges of basketry structures that were used in the early development of pottery to support the walls of the vessel prior to firing. In other cases decorations are remnants of cords or lashing that were once tied around the pots so that they could be carried more easily.

Skeuomorphism is not a thing of the past nor is it limited to traditional crafts. It is found today in countless articles being made from plastic for the first time. Plastic, which can be molded into almost any shape and color, will most often be given a form dictated by the conventional shape of the artifact. The first plastic water pails were patterned after their galvanized steel predecessors. The first plastic baskets appeared in shapes that were originally determined by the reeds and wood splints from which older baskets were fashioned. Only somewhat later did plastic pails and baskets assume forms that were relatively free from the influence of sheet metal and plant material.

Skeuomorphs, routine innovations, and random variations all highlight the conservative aspect of handicrafts. According to design theorist Christopher Alexander, resistance to change is the essence of the traditional crafts and the source of their strength. Alexander identifies two general approaches to the design and making of new artifacts. The first, associated with primitive societies and handmade things, is an *unselfconscious process*. Craft skills are passed on by experienced workers conducting demonstrations and by novices doing a trial-and-error copying of existing artifacts. Because the knowledge of such crafts is not summarized in a written text or put into extended oral form, there are no general theories to be studied. One learns by doing, and what one learns to do is what has been done in the craft for many years, perhaps centuries. These crafts have no place for the individual who hopes to gain attention by personal displays of inventiveness.

Traditional artifacts made by this process have been subjected to many small improvements over the years and are thus admirably suited to their function. Primitive peoples may have few tools or utensils from which to choose – a canoe, an axe, a clay pot – but those that are available are very good ones, given the prevailing level of material culture.

The situation is different in modern societies, where we find the second approach to the design process – a *self-conscious* one in which the novice is presented with a great deal of theoretical knowledge, surrounded by very complex artifacts, and encouraged to see himself or herself as an innovator. The result, Alexander concludes, is a world filled with a bewildering diversity of artifacts, many of which are ill-suited to do the job expected of them.

Alexander's thesis, which owes much to the views of anthropologist General Augustus Henry Pitt-Rivers on primitive technology, is a very engaging one but is not well supported by historical evidence. The examples customarily cited – proving that change is slow and small in traditional societies, that innovation is suppressed, and that their artifacts are functionally superior – have been gathered by observers over a short span of time. We lack extended histories of technology of primitive societies. Perhaps if we had them, we would find counterexamples documenting large-scale innovations in the crafts practiced by primitive peoples.

This is precisely what has been discovered in the development of technology in the Near East during the last five thousand years of the prehistoric era (ca. 8000 to 3000 B.C.). Although prehistoric cultures cannot be equated with modern primitive ones, their technologies do share certain common features: the extensive use of hand tools; a limited employment of machines, and then only the simpler ones; a reliance on human and animal power; and the absence of a scientific basis. Given these restrictions, what can be accomplished in the way of technological innovations? It would seem that a great deal can be done. One can argue that at least as much was achieved in late prehistoric times as during the past five millennia of recorded history.

The late prehistoric epoch witnessed an expansion of the materials inventory available to craftworkers. Older materials such as stone, wood, and bone were now supplemented by copper, bronze, gold, silver, tin, brick, and pottery. These additions required the extensive modification of natural substances (ores and clays) before they were ready for use. Novel techniques for working them also had to be devised, such as the casting of metal and the shaping and firing of clay. These innovations in turn spurred the invention of a whole panoply of tools needed to transform the materials into goods for the community.

Power and transport also underwent considerable advancement in this period. The domestication of animals and the related invention of various kinds of harnessing made it possible to use animal power effectively in agriculture, transportation, and food processing. Innovations in land transport began with the invention of sledges to haul heavy loads and climaxed with the creation of wheeled vehicles and the roads they required. Meanwhile, on rivers and streams there appeared the first rafts, then canoes, and finally boats.

The number of innovations clustered about agriculture and domestic life is overwhelming. The introduction of agriculture augmented and stabilized the food supply, aided the growth of larger settlements, encouraged the development of irrigation systems, and led to the invention of the plow and tools associated with the cultivation, harvesting, and storage of crops. The shape of twentieth-century everyday life owes much to these early agriculturists, who were the first to live in settled dwellings furnished with cooking utensils, textiles, baskets, mats, and furniture, and to make paints, perfumes, soaps, and dyes.

Last in the catalog of late prehistoric accomplishments is writing.

This invention, which joined manual skills with intellect in a manner unique in human history, also required the creation of accompanying physical implements: clay tablets and styli. Writing marks the close of an extremely productive and fertile period. The inventions of this period are significant for two reasons: First, they set the stage for the subsequent rapid growth of Western material culture and civilization; and second, they stand as proof that important, large-scale innovations can occur in a technological setting that would be rated inferior or primitive by modern standards.

Economic Incentives

The emergence of technological novelty in the relatively simple economies of traditional societies is of far less interest to economists and economic historians than is its manifestation in the modern industrial world. There is a growing body of economic literature, both theoretical and empirical, dealing with invention since the late eighteenth century. The contending viewpoints embodied in this literature have helped to promote a broader understanding of the economic dimensions of the inventive process.

Although Karl Marx was not the first to offer an economic explanation of technological change, his work is among the better-known discussions of the subject. Marx readily acknowledged the great technological achievements of industrial capitalism. In subduing nature with the help of steam engines, railways, electric telegraphy, and machines of all sorts, the industrial class within a century managed to surpass, he claimed, the accomplishments of all past civilizations. Egyptian pyramids, Roman aqueducts, and Gothic cathedrals were no match for the monuments of modern industrialism. The capitalists were spectacularly successful because they were members of the first ruling class in history to reject a static society and identify themselves with a dynamic one driven by unceasing technological change. As Marx wrote in the *Communist Manifesto*, "Constant revolutionizing of production, uninterrupted disturbance of all social conditions, everlasting uncertainty and agitation distinguish the bourgeois epoch from all earlier ones."[2]

The capitalist's frenzied pursuit of change was an effort to increase profits, enlarge markets for manufactured goods, and maintain control over the men and women employed in factories. The last reason was an especially important one. "It would be possible," said Marx in *Capital* (1867), "to write quite a history of the inventions, made since 1830, for the sole purpose of supplying capital with weapons against the revolts of the working-class."[3] He had in mind the technological innovations deliberately made to thwart recalcitrant or striking workers. Marx's call for a history of the impact of industrial conflict upon technological change has yet to result in a full-scale study of the subject; however, a beginning has been made by Tine Bruland who has linked three key inventions in the nineteenth-century British textile industry to chronic labor problems.

The first invention is the self-acting, or automatic, spinning mule (1824). This machine spun cotton thread without the aid of workers, except for a few who were needed to repair broken threads and to lubricate and maintain it. Earlier, nonautomatic mules required the attendance of skilled and highly paid operatives called spinners. The spinners accounted for 10 percent of the work force in a cotton mill yet they were absolutely essential to its operation. They used their crucial position to acquire quasi-managerial powers, dictate working conditions, and obtain pay raises. Cotton manufacturers, resenting the control that spinners exercised over production, sought the help of inventors in the creation of a self-acting mule. Richard Roberts was the first to complete a successful machine and did so following a three-month-long strike that had closed the mills at Hyde, England. Although the automatic mule did not result in the immediate industrywide dismissal of spinners, its existence diminished their independence, depressed their wages, and limited their propensity to strike.

The second of the textile inventions, the cylinder printing machine, revolutionized the printing of calicoes. Traditionally, printers used engraved blocks of wood, measuring five by ten inches, to impress designs on calico cloth. The production rate was slow. Printing a twenty-eight-yard-long piece of cloth required the application of the inked block, by hand, 448 times. These skilled printers were members of an old and well-organized trade union. After a series of printers' strikes in the late eighteenth century, mechanical textile printing was introduced. The hand held block was replaced by a cylindrical metal roller on which the design was engraved. Cloth fed into the inked roller was printed quickly and accurately. The hand-block printers lost their power rapidly as more and more manufacturers adopted the machine.

The mechanization of wool combing is the third of the textile inventions associated with labor conflict. Before wool could be spun into thread, its tangled fibers had to be aligned in parallel strands. This was done by wool combers using heated hand combs, work that was arduous and required special skills. Wool combers, like calico printers, were members of an established trade union, and they were noted for their independence and militancy. In fact, they were so powerful that acts of Parliament were passed early in the eighteenth century to curb their influence in industry. Because of technical difficulties, wool-combing machines were slow to be perfected. The first such machines appeared in 1790 and their further development was accelerated by strikes of combers in the 1820s and 1830s. By mid-century effective wool-combing machines were being built and the combers began a battle of resistance that was doomed to failure.

Although the strength and autonomy of wool combers, calico printers, and spinners led textile manufacturers and inventors to devise machines that would replace them, industrial strife is but one of the many economic factors that have been suggested as spurs to invention. The claim has been made that periods of heightened commercial activity are invariably times of increased inventiveness. At those times excess profits are available to be spent on technological novelties that were hitherto considered financially risky. A contrary school of thought argues that during slack periods, when the economy is in a slump, inventors seek the technological improvements that might change the situation for the better. If we fail to find evidence of increased inventiveness during economic depressions, it is because the innovations fostered then are not exploited until economic conditions improve. Related to these interpretations are attempts to correlate increased inventive activity with changes in long-term business cycles or in the general history of prices. One economist has argued that any price change acts as a double stimulus to invention by creating cost difficulties in one field and the opportunities for profit making in another one. The inventor, aware of this situation, shapes his innovation to take advantage of existing constraints and benefits.

The recent awareness of the imminent shortage of crucial raw materials for industry has drawn attention to the ways such shortages were dealt with in the past. One response of modern, technologically versatile societies to material shortages is technological innovation. Economic historian Nathan Rosenberg has listed several of the options available to industrial societies when faced with the prospect of a diminishing supply of a natural resource. One possibility is to raise, through technological improvements, the unit output of the critical substance. Coal, for instance, has long served as a fuel for the generation of steam power. In the twentieth century when coal, which was increasingly used in the production of electricity, became more expensive, the efficiency of steam electric-generating plants was improved to the point that a kilowatt-hour of electricity in 1960 required only 0.9 pounds of coal, as compared to 7 pounds in 1900.

Another innovative technological solution to a material shortage, outright substitution, may result in the invention of wholly new materials, such as synthetic fibers or plastics, or the alteration of existing technologies to accommodate a substituted natural material. The substitution of coal for wood in England is a notable example of the latter. As early as the sixteenth century, laws were passed protecting English timber as a scarce resource. In preindustrial times wood was used as a building material, as a fuel, and, in the form of charcoal, as an essential ingredient in the making of iron. For more than a century wood was slowly replaced by coal as one industry after another made the necessary technological changes, drastic ones in some cases, enabling them to utilize the more plentiful energy source. Many of the important inventions of the Industrial Revolution, especially in iron production, resulted from this substitution.

Of all the economic inducements to innovations, two have received special attention from economists and historians, and have generated wide discussion: the role of *market demand* in innovation and the stimulus to invention provided by *labor scarcity*.

Market Demand

In his very influential book, *Invention and Economic Growth*, Jacob Schmookler claims that inventors draw upon preexisting scientific and technological knowledge as they shape their inventions to satisfy some human want or need. Inventions, therefore, are fusions of an intellectual past with a socioeconomic, functional future. The crucial question is whether inventions are stimulated by the push of the growing supply of knowledge or by the pull of the increasing demands of the marketplace. Schmookler marshals a large amount of data to support his case that market forces are the most important ones.

A historian might be satisfied with the relatively few inventions mentioned in histories of technology, but an economist, seeking information on numerous inventions over an extended time, must look elsewhere. Patents serve as indexes of the number of inventions produced for private industry in different fields and over different time periods, so Schmookler chose American patent statistics as his main source of evidence. These data are difficult to interpret, yet Schmookler was convinced by recent studies that a high proportion of patented inventions (about 50 percent) found commercial application and that there was no other comparable source of data on inventions.

Schmookler first dealt with the supply-side theory of invention, which viewed the growth of scientific and technological knowledge as the force that drove inventive activity. His study of patents led him to conclude that the body of scientific knowledge may shape the overall course of inventive activity, but it is not responsible for the appearance of individual inventions.

If knowledge is not the causal agent, then it might be that the chain of inventions is a self-sustaining one with earlier inventions stimulating later ones. Sociologist William F. Ogburn and others had long claimed that technology grew exponentially, that the accumulation of inventions served to inspire more inventiveness, which led to an even greater stockpile of inventions and hence more inventive activity. Schmookler looked for confirmation of this theory in his patent records but failed to find it.

Having exposed the inadequacies of prevailing theories of invention, Schmookler hypothesized that inventive activity is governed by the expected value of the solution to technical problems. Inventors are moved to contrive novel solutions to technical problems when a financial reward is possible; the higher the likelihood of a reward, the greater the number of solutions, or inventions, produced. Schmookler then tested the validity of this hypothesis by subjecting it to analysis using patent statistics and relevant economic data.

Schmookler chose a select group of inventions for intensive study, those made in the capital goods sector of the economy. Capital goods, or goods used in the production of other goods, include machinery, factory equipment, buildings, locomotives, and trucks. By establishing a correlation between capital-goods invention and capital-goods investment, Schmookler essentially confirmed his demand theory of invention. His reasoning ran as follows. When an industry invests heavily in capital goods, installing new assembly lines or replacing old production equipment, inventors are motivated to produce inventions for that industry knowing that they will be rewarded for their efforts. Therefore, the pull of a particular capital-goods market induces inventors to create new machines and devices.

A close inspection of Schmookler's data linking capital-goods investment and invention presents proof that a causal relationship exists between the two. In general, inventions peak when investments peak and decline when investments decline, but the correlation does not always hold. A time lag exists between the troughs of the two, with inventions temporarily lagging behind in most instances. The invention time lag is important for Schmookler's theory because it tends to support his claim that inventors respond to variations in investment. Had investment responded to an earlier spate of invention, then the supply-side explanation of invention would have triumphed.

Although Schmookler's book is a pioneering study based on the novel and imaginative use of patent statistics, it contains some major weaknesses. An obvious shortcoming is the restriction of the demand-pull theory to capital-goods invention. Neither Schmookler nor any other economist has offered a fully articulated, market-demand theory of inventions for consumer goods such as automobiles, home appliances, or prepared foods.

Another, and possibly more serious, flaw exists in Schmookler's methodological approach, specifically his equation of patent with invention, his assumption that patents provide a reliable measurement of inventive activity. By treating patents as if they were so many equal units of inventiveness, he obscures the real distinctions that separate them. Some inventions form the basis of entirely new industries or radically change existing technologies; others provide small improvements in minor devices; and the remaining, and largest, group have little or no economic impact.

Schmookler's strong reliance on demand-side explanations and his rejection of supply-side, knowledge-induced inventions weakened the structure of his overall argument. By stressing the level of investment and the number of patents (inventions), he was led to claim that demand could solve any technical problem as long as it was profitable to do so. This claim can be refuted by a consideration of the problems unresolved in our day. The need for cheap and nonpolluting energy sources, pest-resistant grains and trees, or a cure for cancer is clearly evident. Inventions that would satisfy these needs would be highly desirable and profitable. Nevertheless, they are not being turned out in regular succession.

Labor Scarcity

In the science of economics the proposition that a scarcity of labor induces a search for labor-saving inventions was first formulated by John R. Hicks in 1932. Hicks argued that capital goods inventions "naturally" incline toward a reduction of whatever factor of production, capital, or labor that shows signs of becoming scarce. Because capital was more readily available than labor in Europe over the past several centuries, he concluded that there existed a natural stimulus to the creation of labor-saving inventions. An earlier version of Hicks's idea can be found in the writings of mid-nineteenth century travelers, engineers, and manufacturers who compared industrial developments in Britain and America. These observers attributed the preponderance of labor-saving inventions in the United States to the scarcity of labor there.

"The Great Exhibition of the Works of Industry of all Nations,

1851" celebrated England's position as the workshop of the world; however, in the midst of the many superior British manufactured goods were some American-made products that caught the public's attention. These items represented a departure from existing British and European manufacturing practices and pointed to a new, American way of making things. This distinctive mode of production, known as the "American system of manufacturing," made extensive use of special-purpose, labor-saving machines that were operated sequentially to shape the components that made up the finished products.

A century after the Great Exhibition, economic historian H. J. Habakkuk published a comparative study of labor-saving inventions used in early nineteenth-century America and Britain. He took a subject that hitherto had been loosely discussed by commentators on the economic and social scene and placed it within the context of modern economic theory.

Habakkuk opened his study by calling attention to the abundance of fertile and accessible land in the United States during the first half of the nineteenth century. Agricultural output was high and profits went directly to the farmers who owned and worked the land. American industry was thus forced to offer wages that were competitive with those available in agriculture. An abundance of land and a shortage of labor drove Americans to produce labor-saving inventions for agriculture, as well as for industry. The McCormick reaper is the most famous but by no means the only example of a machine that made it possible for the American farmer to cultivate more land with fewer hired hands. Mechanization was a much less attractive alternative for the agriculturist in Britain where land was scarce and labor abundant and cheap.

Industrial labor in America was not only dearer than it was in Britain, its supply was also less elastic – that is, American industry as a whole found it difficult to get additional labor when it was needed. The abundance of land, the relative sparsity of the population, and the high cost of transportation contributed to this inelasticity. In eighteenth-century Britain, when labor was more dear and less elastic than a century later, industry had responded by adopting labor-saving techniques. This response led to the technological changes we associate with the Industrial Revolution. America's pursuit of labor-saving inventions in the nineteenth century repeated the pattern.

Before considering the inducements to mechanization operating in America, Habakkuk turned to the question of the relative proportion of skilled to unskilled labor. This matter is critical because

skilled workers are machine makers and unskilled the machine tend-
ers. Without skilled labor there would be no labor-saving machines.
Although skilled workers were dearer than unskilled workers in both
Britain and America, Habakkuk demonstrated that in the United
States the increased demand for labor raised the wages of unskilled
workers more than it did of skilled ones. The supply of skilled labor
was, relatively speaking, more abundant than unskilled labor in
America. Therefore, when the demand for American industrial labor
rose, skilled workers were available at wages that made it possible
to use them to fashion and invent the machines that would replace
the scarce unskilled labor force.

Most nineteenth-century observers agreed that it was the dearness
and the inelasticity of American labor that induced entrepreneurs
to replace labor with machines. In the language of economics, scarcity of labor led Americans to use techniques that were capital
intensive. In times of economic expansion, when the supply of
unskilled labor was especially low and the supply of capital high,
the relative abundance of skilled labor made it reasonable to seek
capital-intensive, labor-saving methods of production.

The entrepreneurial decision to mechanize posed no threat to
American common laborers who were in short supply. In Britain,
however, the introduction of new labor-saving machines meant that
either employed unskilled workers might be laid off or that the
unemployed would find it more difficult to get a job. Whereas
American workers accepted labor-saving inventions, their British
counterparts resisted them through strikes and the destruction of
the machines that threatened their livelihood.

Habakkuk points out that the tradition of American labor-saving
inventions began impressively in the late eighteenth century. At
that time, machines were patented for the manufacture of nails and
pins and Oliver Evans invented his automatic flour mill. Evans did
not alter the methods by which grain was ground into flour; he
revolutionized the movement of grain within the mill. Instead of
using manual power to lift, load, and move grain, Evans devised
water-powered mechanical conveyances that automatically transported grain to the appropriate machine or work station without
human intervention. This reduced the work force needed to operate
a mill by 50 percent.

The stream of American labor-saving inventions continued un-
checked into the early decades of the nineteenth century. In 1841 a
witness before a British Parliamentary Committee investigating the
export of machinery was adamant: "The chief part, or a majority, at
all events, of the really new inventions . . . have originated abroad es-

Figure IV.2. The Blanchard lathe (1820), a special-purpose woodworking machine invented in America. It was capable of reproducing gunstocks, shoe lasts, axe handles, and other irregularly shaped wooden objects. In this illustration a model shoe last (T) is being reproduced at U from a rough piece of wood. Source: Edward W. Byrn, *The progress of invention in the nineteenth century* (New York, 1900), p. 369.

pecially in America."[4] American inventors were even surpassing their British rivals in improving the textile machines that had earlier initiated industrialization.   Foremost among industrial innovations was the novel, "American system," way of manufacturing goods, which relied extensively on machines for the working of metal and wood. Machine tools – the general-purpose drills, lathes, and planers used in shaping metal – had been invented by the British, but Americans adapted them to special-purpose or single-purpose use in their factories (Figure IV.2). Then they enlarged the scope and versatility of these tools by inventing the turret lathe and the milling machine.

The high degree of mechanization that marked the American system powered the drive for the production of standardized, interchangeable components that U.S. government arms-making establishments used. The result was that a radically new manufacturing technology emerged in industrial America by the end of the nineteenth century. Following the early successes in the making of firearms, it spread to the manufacture of sewing machines, typewriters, and bicycles before it was transformed into the mass production of automobiles and household appliances in the twentieth century. The origins of this novel method of production can ultimately be traced to the nineteenth-century labor shortages that motivated American manufacturers to develop and utilize labor-saving machinery.

Habakkuk's influential account of Anglo-American industrial differences has its supporters as well as a fair number of detractors. His critics ask why American businessmen, faced with higher costs, should have focused exclusively on labor-saving inventions for relief when they could have sought other remedies, including the use of abundant and cheap natural resources.

Habakkuk's work remains a provocative and controversial attempt to interpret early American technological history by the application of economic theory to historical events and data. Whatever Habakkuk's shortcomings, he shares with Schmookler the distinction of directing attention to one of the main ways economic forces may generate technological novelties in modern times. Habakkuk and Schmookler can be criticized for laying too great a stress upon the economic basis of innovation. Rather than dismiss their work for this defect, however, it might be possible to incorporate their insights and conclusions into a more inclusive theory of technological change.

Patents

During the past four hundred years, Western societies have developed economic incentives to encourage technological change. They have used monetary awards as a spur to innovation, passed laws to protect the inventors' rights to exploit the inventions, and created special institutions where innovators were employed to work undisturbed on their projects.

Prizes of money for inventions have the advantage that they can be directed to a special problem, dramatizing the urgency of its resolution. They are far less useful for systematically stimulating technological ingenuity and seldom offer legal protection to the creator of novelties. These goals call for the intervention of the state to create institutions such as patent offices and draft legislation to provide patent laws.

Monopolies by patent from the Crown were first granted in the late Middle Ages and the early Renaissance. These guaranteed the holder's right to gain a financial advantage through the control of a consumer product, exploration of new territory, or development of an invention. Patents bestowed according to the wishes and whims of a monarch gave way in the eighteenth and nineteenth centuries to patents shaped by democratic and industrial forces. These modern patents, often viewed as stimuli to technological progress, granted limited monopolies to inventors thus enabling them to exploit their creations for profit.

The American patent system, which drew heavily on English precedents, was established in 1790. Some of the authors of the Constitution thought that prizes or premiums would suffice but a patent board was created and authorized "to grant patents for any such useful art, manufacture, engine, machine, or device as they should deem sufficiently useful and important."[5] The secretaries of state and war and the U.S. attorney general, who constituted the board, evaluated about fifty patents by 1793. In that year the law was amended so that the courts, and not the patent board, determined patentability.

Changes in patent law and practice continued into the nineteenth and twentieth centuries as interested Americans addressed a series of thorny questions: Who is to judge if an invention is truly novel, useful, or important? On what grounds should these judgments be made? Should we accept the inventor's word on originality in evaluating an invention? Are patents inherently elitist, monopolistic, and therefore antidemocratic? Are some discoveries – scientific laws, mathematical theorems – unpatentable?

These questions were broached, if not put to rest, by the patent office with its files, examiners, and bureaucracy; by extensive legislation and patent lawyers who provided legal advice to inventors; and by litigation that grew in intensity as corporation battled corporation for control of patents that would give them the exclusive right to lucrative markets.

The patent system in the United States, like that in other Western countries, inspires some popular and often-articulated notions that have rarely been subjected to close scrutiny. It is commonly believed that patents promote technological ingenuity, enrich the economies of nations, offer a proper measure of the technological and economic state of a society, and compensate deserving, creative individuals for their hard work. However, analytical and historical examinations of patents and their meaning for technology and economic growth are scarce. Until such studies are made we must be cautious about accepting at face value the truisms found in the many flattering assessments of our patent system.

Economic analysis does not support the claim that economic growth and patenting are obviously and closely linked. If, for example, patent activity is measured in relationship to the growth of the gross national product (GNP), in the twentieth century one finds a great discrepancy between the two. Since 1930 the GNP has maintained a lead far in advance of the increase in patented inventions. Evidence of this sort supports the conclusion of the distinguished economist Fritz Machlup in his study of the economic impact of the patent system: "No economist, on the basis of present knowledge, could possibly state with certainty that the patent system, as it now operates, confers a net benefit or a net loss on society."[6] This observation, made in 1958, still holds true today.

One source of difficulty encountered in evaluating the modern patent system is that ambiguous person, "the inventor." The name conjures up the image of a struggling, lone figure who deserves rewards for the effort spent, and the risks taken, in bringing humanity a new and useful thing. Yet, since the late nineteenth century, an increasing number of inventors have been working in industry and not in the confines of their home workshops. Although patents can only be issued to individuals, not corporations or institutions, those inventors who work in industry are, as a condition of employment, routinely required to assign their patents to their corporate employers.

The radical shift to corporate ownership of patents has had a great impact on the social welfare and economic life of the American people. The monopoly that the founding fathers bestowed on a private individual has passed to large, powerful, and wealthy corporations who are able to control entire industries through the purchase and manipulation of patents. The corporate inventor, who contracted to exchange inventions for job security, made it possible for the corporation to become, in effect, the inventor and to acquire unprecedented monopolistic rights.

Originally the seventeen-year monopoly was meant to protect inventors as they prepared their inventions for market. Once the corporation gained control of patents, the monopoly was used to suppress any inventions that might harm its own products or enhance those of a rival. Furthermore, corporations use their captive inventors to devise machines and processes that protect and perpetuate their own patented products and encroach upon those of competitors. Social benefits are rarely the concerns of those involved in these maneuvers.

Criticism of the patent system implies that there are better ways of encouraging inventive activity and ensuring the welfare of inventors and society. However, because the patent system and modern industrialism emerged simultaneously during the Industrial Revolution, we have had little experience with alternatives to patents.

British economists C. T. Taylor and Z. A. Silberston compensated for the lack of alternatives by comparing the British patent system of the 1970s with a hypothetical alternative. What would be the impact on the British economy, they asked, if the monopoly element of the patent were removed? In place of the sixteen-year monopoly granted to British inventors, a patent would be issued with the requirement that its holder accept any legitimate request for a license to use the invention. The licensing fee could not be set so high that it would create a monopolistic situation, and if necessary an official arbitrator would be called in to fix licensing fees and royalty payments at a rate agreeable to all parties.

With the compulsory licensing system so defined, Taylor and Silberston contacted a select number of British industries requesting them to assess the proposed alternative in the light of current patent law and practice. The results of their inquiry may come as a surprise to enthusiastic supporters of the patent system, for they found only a slight economic advantage in keeping the existing patent system; the respondents also indicated that patents on the whole offered a very limited inducement for industrial invention. Research activity in the drug and pesticide industries would be adversely affected by the abolishment of the patent system because those industries rely on monopoly of inventions. Other segments of the chemical industry (plastics, artificial fibers) as well as mechanical engineering and the electronics and electrical industry foresaw little or no difficulty with compulsory licensing. On the other hand, small firms and individual inventors, both of whom depend on limited monopoly for protection against large companies, would be harmed by the end of patenting.

Most Western industrial countries instituted national patent legislation sometime in the nineteenth century and have kept it in force. The Netherlands and Switzerland were exceptions to this rule. For long periods both nations were patentless, although their citizens could patent inventions abroad. The Dutch abolished their inadequate patent system in 1869 and did not replace it until 1912. The Swiss were without patents until 1907.

How did these two countries fare during the high tide of European industrialization without patents to spur inventiveness and ensure industrial progress? Economist Eric Schiff, in a study of the Dutch (1869–1912) and Swiss (1850–1907) economies during the patent-free years, showed that neither were hampered economically by the absence of domestic patenting. Industrial progress in the Netherlands was roughly comparable to that of other European nations and in the case of two Dutch industries, margarine and incandescent light bulb production, the lack of patents acted as a positive stimulus. During the patent-free period the Dutch economy was more dependent on trade than industry but that was the result of peculiarities in the economic development of the Netherlands that antedated the mid-nineteenth century. The Swiss example is more striking. Switzerland experienced vigorous economic growth between 1850 and 1907. Industry was so successful that it attracted foreign capitalists who were willing to invest in new ventures despite the absence of patent protection. Overall the Swiss were more inventive than the Dutch during the patent-free years they shared. After 1912 the rate of domestic Dutch inventions did increase somewhat.

If the Swiss and the Dutch were able to flourish economically without the burden of administering and financing a national patent system, why did they bother to adopt one eventually? Primarily, they were under moral and political pressure from the community of industrial nations that had banded together in the International Union for the Protection of the Industrial Property, an organization dedicated to the mutual protection of the rights of patent and trademark holders.

A convincing test of the efficacy of patenting might be made if we knew more about the situation prevailing in the Communist countries. The exploitation of inventions for profit in those lands was rejected as capitalistic and the state appropriated the monopolistic rights to inventions. Some economists have suggested that the absence of the financial incentives provided by patents accounts for the lackluster performance of state-controlled pharmaceutical research behind the Iron Curtain. The Soviet Union, for example, has contributed no major drugs to the world's pharmacopoeia. Yet factors other than the lack of a traditional patent system may explain the Russian failure to innovate.

A patent of sorts is to be found in the USSR. The *author's certificate* was introduced immediately after the Revolution (1919) and, in the words of the *Great Soviet Encyclopedia*, "confirms the author's claim to authorship and his right to a reward and other rights and benefits, and the government's exclusive right to use the invention."[7] The reward, rights, and benefits are not specified. One assumes that the certificate holder negotiates with the government, or accepts its evaluation of an invention's worth. Eastern bloc countries have followed Russia's lead in dropping patents as capitalistic anachronisms, but they too have replaced them with some other formal legal document recognizing the contributions made by an inventor. It appears that whatever the ruling political ideology, the idea of a reward to the creator of technological novelties has a secure place among modern nations.

The significance of patents is not that they offer strong and indisputable incentives for invention. The most that can be said is that at some times and under certain circumstances patents have probably been beneficial in promoting economic growth and inventiveness. In fact, the effectiveness of the patent system is less important than the fact that every industrialized country in the West has made patenting a national institution, complete with supporting bureaucracy, legislation, and state funding. When combined with the zealous pursuit of patents by industry, the existence of professional careers in patent law practice, the transformation of the patent in Communist countries, the popular enthusiasm for the idea of the patent, and the economist's and historian's interest in probing the meaning of patents, the result is an obsession with technological novelty that is without precedent. No other cultures have been as preoccupied with the cultivation, production, diffusion, and legal control of new machines, tools, devices, and processes as Western culture has been since the eighteenth century.

Industrial Research Laboratories

The pursuit of patents is associated with industrial research laboratories, the first of which was established late in the nineteenth century. Previously, scientists had worked in industry as consultants or were scientist–entrepreneurs who founded their own firms. The opening of research laboratories led to the employment of scientists as salaried research workers and to the industrialization of invention. One of the main reasons for the support of scientific research by industry was the realization that science could be put to work creating patentable innovations that yielded new and improved products.

The first industrial research laboratories were organized in the 1870s and 1880s by synthetic dye manufacturers in Germany. The dye industry, which drew heavily on the advancing research front of organic chemistry, initially (1860s) obtained new dyes by purchasing patent rights from the independent chemists who held them. By the end of the 1870s, producers of dyestuffs saw the advantages of establishing in-house laboratories and hiring full-time chemists to staff them. Chemists could aid in the solution of routine problems encountered in the manufacturing process, but they would be even more valuable in creating new dyes, of different hues and intensities, suitable for coloring a variety of textiles.

These developments take on an added significance when one learns that synthetic dye chemistry did not originate in Germany. The first synthetic dye, aniline purple, was discovered by young English chemist William H. Perkin in 1856. Within a few decades of the discovery of aniline purple, the Germans had not only developed and monopolized the production of synthetic dyes but had revolutionized the relationship between science and technology by employing research scientists to advance industrial goals.

In the United States the electrical industry was the pioneer in industrial research. Thomas A. Edison's private laboratory at Menlo Park, New Jersey, established in 1876, served as an early example of what might be accomplished when organized research was enlisted in the solution of technical problems. The inventor exaggerated when he boasted that he could turn out "a minor invention every ten days and a big thing every six months or so";[8] however, his development of an effective electric incandescent light bulb vindicated the concept that a team of researchers, each with different talents and specialties, could concentrate their efforts on a single problem, while isolated from the distractions of routine production work.

The General Electric Company established the first corporate research facility in America. The company grew out of Edison's technical and business interests, but by 1889 the founder was playing a minor and diminishing role in company affairs and GE was on its way to becoming a major producer of electrical goods ranging from light bulbs to dynamos. A decade later GE was somewhat unsure of its place in the growing electric industry. Key patents that had sustained its earlier growth had expired, and independent inventors who were once associated with GE had either died or gone elsewhere. Determined to maintain its position as a leader in electric lighting at home and abroad, in 1901 the company decided to establish its own research laboratory. One GE executive justified the decision thusly:

Although our engineers have always been liberally supplied with every facility for the development of new and original designs and improvement of existing standards, it has been deemed wise during the past year to establish a laboratory to be devoted exclusively to original research. It is hoped by this means that many profitable fields may be discovered.[9]

Shortly after GE opened its research laboratory other well-known American firms followed suit. In 1902 the Du Pont Company and the Parke-Davis pharmaceutical company founded research laboratories; Bell System formally created its research branch in 1911; and Eastman Kodak established a photographic research laboratory in 1913. The first firms to engage in organized research were those whose technologies were closely linked to two areas of science that flourished in the late nineteenth century, chemistry and electricity.

The number of American laboratories employing scientists and engineers in industrial research has grown rapidly. Within twenty years after the establishment of the GE laboratory, 526 American companies had research facilities. By 1983 that number had risen to 11,000.

Exactly what use does a firm make of its research laboratory? What does industry expect to gain in return for its investment in original research? These questions are answered with least ambiguity in the case of the earliest industrial laboratories. In the beginning research was undertaken to cope with some pressing technical problem. The German dye firm of Bayer and Company opened its research laboratory in the hope of exploiting the newly discovered azo dyes that promised to dominate technology in the future. General Electric, whose high-resistance carbon filament incandescent light bulbs were being threatened by W. Nernst's "glower" lamps and P. Cooper-Hewitt's mercury vapor lamps, decided to protect itself by formally sponsoring research in electrical lighting. The Bell System's laboratory grew out of the need to develop effective long-distance telephony and to respond to the challenge posed by the ongoing experiments with wireless (radio) communication.

The publicly stated reason for corporate financing of research generally has been that new knowledge is almost certain to lead to new, better, and cheaper products. In a sense money spent on research can be thought of as an investment in the long-range profit potential of the firm. This reasoning, which represents the aggressive, or offensive, business strategy associated with industrial research, is supported by modern examples of valuable commercial products that have emerged from corporate research ventures: Nylon and other synthetic fibers, detergents, antiknock gasoline, improved automobile engines, the newer plastics, many modern drugs, television, and transistors.

A defensive strategy for industrial research is equally as important as the offensive one but not as well known. A successful industrial research laboratory is a generator of patents, not all of which will be transformed into commercial products or internal improvements in manufacturing. Patents may prove to be most profitable when used as counters in the struggle between corporate rivals.

A case in point is the relationship between patents and research at the Bell laboratory during the first decades of its existence. The idea of research as a producer of new and useful knowledge was altered as research administrators and Bell executives learned that it was in the company's interest to patent every possible minor variation of a device in order to forestall future encroachment by competitors. Researching new frontiers of technological knowledge might be far less productive than, in the words of a Bell president, occupying a technical field with "a thousand and one little patents and inventions."[10]

The prevention of competition was just one of the new uses Bell, and others, found for patents. Patents also could be acquired to frustrate a competitor's attempt to secure a strong patent position through its own research efforts. In obtaining these critical patents, the first firm often has no intention of competing in the marketplace. The sole aim is to place impediments in the way of a rival's domination of a technical field that is central to its economic success. Finally, patents are often used to provide a strong bargaining chip that, at the right moment, can be traded to a rival for other patents or concessions.

Patents obtained for defensive purposes often remain undeveloped. They exist to provide a shield behind which the corporation can retreat and protect itself from the potential threat of innovating competitors. This defensive strategy makes a very conservative use of industrial research. Instead of serving as a source of innovation, research becomes part of a business maneuver aimed at preserving the status quo, or at least assuring that novelties will emerge at a slowed rate.

In sum, there is a discrepancy between the ideal of industrial research – science stimulating technological change – and its real uses in the world of commerce. A research laboratory may be maintained to bring an aura of science and prestige to a firm, to keep fresh scientific talent on tap within the corporate structure, or to erect a bulwark against change. The generation of technological novelties is one of several functions of the corporate research laboratory but by no means is it the only reason for its existence.

Given the many nonproductive uses to which patenting is put in modern industry (nonproductive in the sense of not appreciably advancing knowledge), the question arises as to how effective the industrial research laboratory is in stimulating innovation. Caught between the conflicting goals of science and commerce, organized research is not nearly as instrumental in creating novel products and processes as corporate self-advertising would sometimes have us believe.

A firm's size and the nature of its technological base influence its role as an innovator. Only the larger firms can afford to fund elaborate in-house research projects. At the same time, these firms are more reluctant to embark on new paths than are smaller and more venturesome companies. Apart from size, firms forced to explore new technological areas – for example, pharmaceuticals or semiconductors – are apt to be more interested in the latest results of scientific research than those associated with older technologies such as automobiles, home appliances, or railroads.

The existence of a large, well-staffed laboratory does not necessarily mean that a firm is self-sufficient in its research needs. The Du Pont Company, a recognized leader in industrial research, maintains extensive laboratories that have been held in high regard by its top officials. In 1950, Du Pont president Crawford H. Greenwalt announced: "I can say categorically that our present size and success have come about through new products and processes that have been developed in our laboratories."[11] Economist W. F. Mueller, who studied the sources of Du Pont innovations during the thirty-year period from 1920 to 1950, came to the opposite conclusion. He found that of twenty-five important new products and processes introduced by the company in that time only ten were based on inventions by Du Pont's research staff. These included five of eighteen new products, and five of seven new processes. The rights to fifteen innovations produced outside the company were obtained from various firms and from independent inventors.

The Du Pont Company figures are indicative of an important fact. The independent inventor was not displaced by the organized research teams that swept into industry after the turn of the century. A study of seventy of the most important inventions produced in the first half of the twentieth century found that more than half of these emerged from the work of independent inventors. The list of their contributions is an impressive one and includes the automatic transmission, Bakelite, the ballpoint pen, Cellophane, the cyclotron, the gyro-compass, insulin, the jet engine, Kodachrome film, magnetic recording, power steering, the safety razor, xerography, the Wankel rotary-piston engine, and the zipper fastener.

Industrial research laboratories are by no means the "invention factories" their promoters claim them to be. Nevertheless, they have offered an alternative career path for scientists and research-oriented engineers, and they have advanced scientific and technological knowledge in fields allied to the corporate goals of their financial patrons. Whatever their uses and purposes, industrial research laboratories continue to enjoy the support of modern business. Like the patent system to which they are closely tied, industrial research laboratories are proof of the willingness of modern industrial societies to invest large amounts of time, effort, money, and materials in order to institutionalize and facilitate the production of novelty.

Novelty and Culture

Discussing economic and institutional incentives to innovation is much easier than dealing with the connections between cultural attitudes and values and artifactual change. Although such connections may appear to be vague and tenuous when compared with the arguments emanating from the economist's camp, they are more useful in explaining why entire societies vigorously engage in technologically innovative activity over long periods of time. The cultural approach is especially relevant to understanding Western domination of the production of technological novelties for the past five hundred years.

As is the case in so many other aspects of modern life, Renaissance culture appears to mark the turning point in attitudes toward the technological innovator. Technologists as a group, whether they were inventors or skilled practitioners, gained greater recognition in the Renaissance than they had in ancient or medieval times. They found patrons to support them and their projects, wrote and published elaborately illustrated books on their technical specialties, and received praise from influential writers and thinkers for the contributions they were making to human welfare.

During the Renaissance, books listing the great inventions and their originators first appeared. Polydore Vergil's *De Inventoribus Rerum* (*On the Inventors of Things*) (1499) was one of the earliest of these popular compilations that sought to identify and honor the inventors of such things as gunpowder, glass, metal, wire, silk, printing, and ships. Sir Francis Bacon took the process one step further in his utopian tale, *New Atlantis* (1627). New Atlantis was a technological paradise with a state-supported research laboratory (Solomon's House) dedicated to the advancement of all of the technical arts. Two large halls in it were set aside to honor the creators of technological novelties: One contained drawings and samples of great inventions and the other statues of their inventors, sculpted from wood, marble, silver, or gold, depending upon the importance of their work.

The recognition bestowed in the Renaissance grew rapidly in the following centuries until with the advent of industrialization inventors became cultural heroes. Books popularizing the lives and

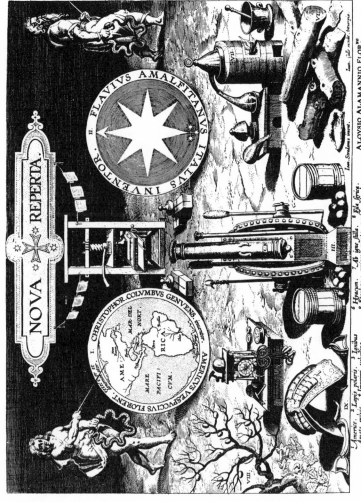

Figure IV.3. One engraving, dating from the 1580s, by Johannes Stradanus celebrates nine great inventions and discoveries of Renaissance Europe: (I) New World, (II) magnetic windrose-compass, (III) gunpowder, (IV) printing press, (V) clock, (VI) guaiacum, used in treating syphilis, (VII) distillation, (VIII) silkworm, and (IX) stirrup, which made armed warfare on horseback possible. Source: *The "new discoveries" of Stradanus* (Norwalk, Conn., 1953).

works of inventors were widely published in the nineteenth century. Bacon's fictional hall of inventions achieved a degree of realization in the international industrial exhibits and museums of science and industry. Statues of inventors were erected in public places along with those of other famous personages, and the government created the patent system to nurture, protect, and reward inventive genius. Twentieth-century society added its own set of honors, including medals of honor presented by government and industry, and well-paying posts in universities, businesses, and governmental institutions.

Recognition and rewards for inventors are now considered to be an appropriate activity to those of us who live in modern Europe or America, but there are other cultures that condemn novelty as strongly as we seek to foster it. In the Muslim tradition, innovation or novelty is automatically assumed to be evil until it can be proved otherwise and applies to innovations made by believers in Islam as well as those imported from other cultures. The Arabic word *bid'a* has the double meaning of *novelty* and *heresy*. The worst kind of bid'a is the imitation of the ways of the infidel for as the Prophet warned: "whoever imitates a people becomes one of them."[12]

Although the West has never in its history universally condemned novelty, the conscious quest for novelty dates to recent times. Historians have traced the origins of the Western craving for novelty to a series of developments that took place in Renaissance Europe (Figure IV.3). Geographical exploration literally discovered new worlds; astronomical observation confirmed the existence of new stars (novae) in heavens hitherto thought to be immutable; medieval scholasticism was replaced by new philosophical systems; and modern science, or the "New Philosophy" as it was then called, presented a revolutionary conception of the universe. By the seventeenth century the fascination with novelty was so great that publishers' book lists were filled with titles promising a *new* alchemy, astronomy, botany, chemistry, geometry, medicine, pharmacopeia, rhetoric, and technology. Of these the best known are Galileo's *Discorso . . . intorno a due nuove scienze* (*Two New Sciences*) (1638), Johannes Kepler's *Astronomia Nova* (1609), and Francis Bacon's *Novum Organon* (*New Logic*) (1620). Lynn Thorndike has concluded that "the new was very much in the consciousness of the men of the seventeenth century."[13]

Closely related to the pursuit of novelty is one of the great and influential ideas of the Western world – the idea of progress. According to its tenets human history does not follow either a cyclic or a declining course; it moves ever onward and upward to a better future. The golden age, therefore, is not a paradise that was lost in past times but one that will be reached in the future. Those who go seeking for wisdom from the ancients must be made to realize that the men and women of the present and future are the true sages. As for the Greeks and Romans, they lived during the infancy of Western culture.

The idea of progress drew its strength from the scientific accomplishments of the seventeenth century that called attention to the cumulative nature of scientific knowledge. Its proponents believed that as modern science assembled its stores of facts and theories, uncovering nature's secrets and gaining control over its resources, humanity would ascend the ladder of progress. All human activity would be transformed by the progressive driving spirit of science.

Technology was a favorite source of examples for the proponents of progress because its implications were thought to be obvious to all. Everyone would agree that the Greeks and the Romans knew nothing about gunpowder and the compass. These recent innovations made by the moderns were a sign of the superiority of the present age and an indication of still greater technological wonders to come.

Francis Bacon was especially fond of contrasting the sterility of speculative philosophy, which had remained essentially unchanged since Aristotle, with the mechanical arts, which had advanced continuously over the centuries. While the scholastics quibbled over minor philosophical points, practical men developed new ways of powering machines, fighting wars, making books, sailing ships, and constructing buildings. These technical changes were unequivocal proofs of progress.

In this context an invention was not novelty for novelty's sake but was one more contribution to the improvement of humankind. Inventions were evidence that humanity was on the road to a better society, perhaps even a perfect one. This optimistic and wonderful vision of the future was severely shaken by the critical arguments of nineteenth-century philosophers, the realities of the less-than-utopian life in factories and industrial cities, and more recently by the horrors of mechanized warfare in worldwide conflicts. Yet that vision endures today. It resides in the hopes of those who are convinced that nuclear or solar energy, space colonies, computers, robots, or biotechnology will surely bring us to the brink of a sublimely happy age.

The domination of nature joined novelty and progress to form a triad of ideas that emerged in the culture of Renaissance Europe and became instrumental in stimulating technological change. The idea that nature exists solely for human use is first found in the account of the Creation in Genesis. God, having given Adam and Eve dominion over every plant and animal, commanded them to subdue the earth and fill it with their progeny. Unlike the Eastern religions in which nature and humanity coexisted on equal terms, Judaism and Christianity established a hierarchy. The creatures made in God's image were placed in charge of the rest of his creation.

Medieval historian Lynn White, Jr., claims that the spectacular success of the West in cultivating science and technology is rooted in the Judeo-Christian belief that the domination of nature was sanctioned by religion. The persistent, aggressive effort made by Westerners to exploit every possible natural force and resource resulted in their becoming the world's leaders in technology. Those whose religions taught them to take a more benign attitude toward nature failed to develop technology to its full potential. The Judeo-Christian viewpoint was augmented and elaborated upon in the seventeenth century by philosophers and essayists who, in Bacon's words, held that nature should be made "to serve the business and conveniences of man."[14] Thus, modern science, which provided a superior means of understanding the natural world, would ensure that it would be mastered more thoroughly.

That nature was to be made subservient to human needs was a topic worthy of commentary and discussion in the seventeenth and eighteenth centuries. By the nineteenth century, the concept was so widely accepted that it could be reduced to a formula expressed in the phrase, the conquest of nature. The idea that technology was to be utilized to its full extent in controlling nature went largely unchallenged until the second half of the twentieth century. Its truth was then questioned by the leaders of the environmental movement who, beginning in the 1960s, argued that in ruthlessly subduing nature we were not only poisoning our surroundings and depleting nonrenewable natural resources but were engaged in an immoral act. What right had we to despoil a realm that predated human life on earth and had its own integrity and goals?

Whether the environmentalists will have a lasting influence on how we think about, and act toward, the natural environment is still uncertain. From the vantage point of the late twentieth century, however, their impact appears limited. A majority of the people living in the Western nations, and many dwelling elsewhere, continue to believe that we should encourage any and all technological novelty because it contributes to the progress of humanity and allows us to carry on the struggle to dominate nature.

Conclusion

Two general conclusions can be drawn from the material presented in this and the preceding chapter. First, there is no broad theory of technological innovation that includes a majority of the factors influencing the emergence of novelty. There are good reasons for the absence of such a theory, because it would have to encompass the irrationality of the playful and fantastic, the rationality of the scientific, the materialism of the economic, and the diversity of the social and cultural.

The second conclusion is that the lack of a satisfactory theoretical approach to novelty does not affect the evolutionary theory presented in this book. That theory calls for an adequate supply of novel artifacts, or ideas of novel artifacts, from which a selection can be made. Chapters III and IV have provided ample evidence for the existence of rich and varied sources of novelty. They have demonstrated that novelty is present wherever, whenever, and by whatever means human beings choose to make things.

CHAPTER V

Selection (1): Economic and Military Factors

Introduction

Because there is an excess of technological novelty and consequently not a close fit between invention and wants or needs, a process of selection must take place in which some innovations are developed and incorporated into a culture while others are rejected. Those that are chosen will be replicated, join the stream of made things, and serve as antecedents for a new generation of variant artifacts. Rejected novelties have little chance of influencing the future shape of the made world unless a deliberate effort is made to bring them back into the stream (Figure V.1).

If these remarks recall the notion of evolution by natural selection, they are meant to do so. There are, however, crucial differences between artifactual and organic evolution that must be mentioned before any further use is made of the selection analogy.

Central to organic evolution is the variability that arises from mutation and from the recombination of the parental genes in sexual reproduction. The resulting variant offspring are subject to natural selection, which permits some, but not all, of the variants to survive, reproduce, and pass on their genetic information. The offspring, which have the potential to move in many different evolutionary directions, are selected by the totality of conditions – environmental, biological, social – that prevail at the time of their appearance. Those that persist have a survival value that is determined by the circumstances they happen to meet and not by any absolute criterion of superiority. Therefore, evolution by natural selection has no preordained goal, purpose, or direction. This is not true for artificial selection as practiced by animal and plant breeders. Here criteria are established by the humans who select characteristics they consider

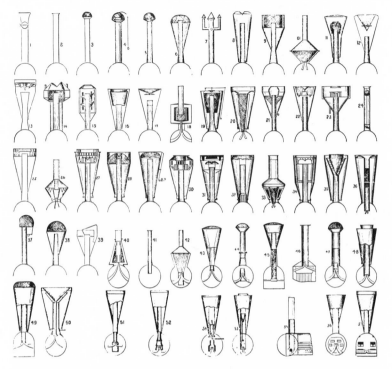

Figure V.1. Excess of novelty: American smokestack spark arrestors (1831–57). The smokestack designs depicted here are fifty-seven of the more than one thousand that were patented in the nineteenth century. Mechanics, inventors, ordinary citizens, and cranks joined in the futile search for a smokestack that would eliminate the escape of sparks and embers from the fireboxes of woodburning locomotives. Source: John H. White, Jr., *American locomotives* (Baltimore, 1968), p. 115.

worthy of preservation: fleetness in race horses, high milk production in dairy cattle, disease resistance in wheat.

Even this abbreviated account reveals important distinctions between the evolution of artifacts and organic evolution. In many respects technological evolution has a great deal in common with artificial selection. Variant artifacts do not arise from the chance recombination of certain crucial constituent parts but are the result of a conscious process in which human judgment and taste are exercised in the pursuit of some biological, technological, psychological, social, economic, or cultural goal. There are, of course, cases in which an artifact changes slowly over time through the accretion of small, barely recognized variations. Stone-tool development may be such a case. For the most part, however, artifactual changes are brought about by men and women using intelligence, imagination, and power to fashion new kinds of things.

From the vast pool of human-designed variant artifacts, a few are selected to become part of the material life of society. In nature it is the ability of the species to survive that counts – the fact that the organism, and especially its kind, can thrive and reproduce in the world in which it finds itself. The artifact may also be said to survive and pass on its form to subsequent generations of made things. This process requires the intervention of human intermediaries who select the artifact for replication in workshop or factory. Because of the complexity and uncertainty of the selection process, survival value becomes an amorphous concept when applied to technology.

In organic evolution the factors responsible for the creation of variants – mutation and recombination – are not the same ones that determine the survival and perpetuation of the species. Here too, artifactual and organic evolution diverge, because many of the forces that encourage the creation of variant artifacts are also influences during the selection process. Chapter IV revealed that the belief in the idea of progress stimulates the invention of novel artifacts. That belief also influences the subsequent selection of the novelty for development by creating a cultural milieu in which new things are welcomed as a sign of betterment.

Finally, variant offspring may, given time, develop into identifiable varieties of existing species or into entirely new and separate species. A biological species may be roughly defined as a group of morphologically similar individuals that interbreed under normal circumstances. The concept of morphological similarity can readily be transferred to made things. The classification of artifacts into different types based on their form and structure, and arranged according to degrees of similarity, presents no insuperable difficulties. In Chapter I we learned that taxonomic work of this sort was begun in the nineteenth century by Pitt-Rivers and others. We do encounter problems, however, when we attempt to apply the notions of interbreeding and fertility to the made world. Different biological species usually do not interbreed, and on the rare occasions when they do their offspring are infertile. Artifactual types, on the other hand, are routinely combined to produce new and fruitful entities.

Anthropologist Alfred L. Kroeber illustrated this critical difference between living and made things by sketching two "family trees": one of organic species, the other of cultural artifacts (Figure

Figure V.2. Family trees as depicted by anthropologist Alfred L. Kroeber. On the left is the tree of organic life; on the right is the tree of cultural artifacts. Source: Alfred L. Kroeber, *Anthropology* (New York, 1948), p. 260; copyright 1923, 1948, by Harcourt Brace Jovanovich, Inc.; renewed 1951 by A. L. Kroeber. Reprinted by permission of the publisher.

V.2). Kroeber's tree of organic life consists of separate branches that split to form new species. The branches remain totally isolated from one another; they never curve and then merge with some other branch (species) to produce novel life forms. In short, this tree looks very much like an ordinary tree. By contrast, the artifactual tree is a bizarre arboreal specimen. Separate types or branches fuse together to produce new types, which merge once again with still other branches. For example, the internal combustion engine branch was joined with that of the bicycle and horse-drawn carriage to create the automobile branch, which in turn merged with the dray wagon to produce the motor truck.

The fundamental differences between the two "trees" show that we must not forcibly extend every element of the biological species idea into the realm of technology. We may fairly assert that technological novelties are selected for replication, and we need not establish which of these new kinds of things are to be designated as a distinct "species" or "type." I therefore use the evolutionary analogy because of its metaphorical and heuristic power and caution against any literal applications, not the least, the process of speciation.

Although we have emphasized these differences, we must not lose sight of the benefits to be gained by using a comparative approach.

On the most general level the evolutionary analogy serves as a useful organizing principle for studying technological change. Up to this point it has enabled us to examine closely the parallel worlds of organisms and artifacts. In doing so we discovered that these two realms exhibit a rich diversity of types and a continuity based on antecedent related forms. We also found that organisms and artifacts share a tendency to produce an excess of novelty, reproduce by means of copying with variations, and spread their innovations over a wide geographical area.

In this chapter and the following one the evolutionary analogy will be used to explore the economic, military, social, and cultural factors involved in the selection of novel artifacts. During the process of selection, humankind is constantly defining and redefining itself and its cultural situation. As it establishes its changing goals, technological choices are made that may affect the welfare of generations to come. This selection process is of crucial importance for present and future human history, yet it does not function in a rational, systematic, or democratic manner. Trial and error predominates as a method, and the small number of men and women who participate in it are subjected to, among other things, economic constraints, military demands, ideological pressures, political manipulation, and the power of cultural values, fashions, and fads. A process open to the influences of such diverse and conflicting forces is not one whose operation is likely to be readily summarized or neatly reduced to a theoretical model. It is best examined through representative examples illustrating how the choice of variant artifacts actually has been made in the past.

General Considerations

Before we review the leading factors affecting selection, some general observations are offered on inventions and the process by which they are chosen and converted into economic and cultural products. These remarks hold true for most of the cases studied in depth in this and the next chapter.

First, the potential, as well as the immediate, uses of an invention are by no means self-evident. Often it is difficult to determine precisely what is to be done with a new device. This was the problem Thomas Edison faced after he had invented the phonograph (1877). The following year he published an article specifying ten ways in which the invention might prove useful to the public. He suggested that it be employed to take dictation without the aid of a stenographer; provide "talking books" for the blind; teach public speaking;

reproduce music; preserve important family sayings, reminiscences, and the last words of the dying; create new sounds for music boxes and musical toys; produce clocks capable of announcing the time and a message; preserve the exact pronunciation of foreign languages; teach spelling and other rote material; and record telephone calls. This listing is important because it represents Edison's own order of priority for the potential uses of his talking machine. Music reproduction is ranked in fourth place because Edison felt that it was a trivial use of his invention. A decade later when the inventor entered the phonograph business, he still resisted efforts to market the phonograph as a musical instrument and concentrated upon selling it as a dictating machine. Others, who saw the entertainment possibilities of Edison's invention, modified phonographs to play popular musical selections automatically on the deposit of a coin. The coin-operated machines, displayed in public places, soon gained popularity. In 1891 Edison was unwilling to accept these early jukeboxes, because he believed they detracted from the legitimate employment of the phonograph in offices.

The successful commercialization and widespread utilization of the phonograph occurred only after it was publicly presented as an instrument for the reproduction of music. By the mid 1890s even Edison agreed that the primary use of his talking machine was in the area of amusement. This led to the establishment of the lucrative phonographic record business that supplied mass audiences around the world with recorded music and reproducing equipment.

One might suppose that the example of the phonograph served as a ready guide to the marketing of the tape recorder when it was first offered to the public shortly after the end of World War II. But that is not the case. The tape recorder, developed in Germany during the war, first came to the attention of Japanese engineers in the late forties. By 1950 Tokyo Telecommunications, the company later known as Sony, was ready to market its own version of the machine, a heavy, bulky, and expensive model. The main problem was to find a use for it that would make it appealing to Japanese consumers. Initially some sets were sold to the Ministry of Justice for recording court proceedings and others to scientists who used them to record data, but sales remained very small until Sony persuaded Japanese schools, colleges, and universities to purchase the machines for the teaching of languages. Even then the market was modest. Not until the 1960s was the tape recorder finally advertised and sold as a device for recording and playing music, at which point sales began to soar.

The early histories of the phonograph and tape recorder have not been recounted to show, with the benefit of hindsight, how blind Edison and the Sony management were to the "true" potential of their machines. More legitimate and relevant points can be made. Clearly the phonograph and tape recorder were not developed to meet some identifiable, pervasive, and pressing need or want. When these machines appeared on the scene, neither the technologists nor the public knew what to do with them. Of course they both reproduced sounds, but to have reached that point in an understanding of them is not to have accomplished much. What sounds were to be recorded? In what social and cultural contexts? There are many possible uses for a new machine, even one as simple as a tape recorder. In fact, as the Sony managers struggled to sell their first models, they obtained and translated an American pamphlet entitled "999 Uses of the Tape Recorder." This advertisement of the versatility of the machine was a sign of weakness, not strength, reflecting the American manufacturers' uncertainty about the product's place in the market.

When an invention is selected for development, we cannot assume that the initial choice is a unique and obvious one dictated by the nature of the artifact. Each invention offers a spectrum of opportunities, only a few of which will ever be developed during its lifetime. The first uses are not always the ones for which the invention will eventually become best known. The earliest steam engines pumped water from mines, the first commercial use of radio was for the sending of coded wireless messages between ships at sea and from ships to shore, and the first electronic digital computer was designed to calculate firing tables for the guns of the United States Army.

The second observation to be made with respect to inventions is that even when there is general agreement about how one is to be used, we cannot assume that it will operate as promised. Initially inventions are likely to be very crude models embodying new ideas in need of further refinement. What decision makers often base a selection on is not a fully developed steam locomotive or transistor but the first working prototypes.

Edison's 1877 talking machine, with its tinfoil recording surface and hand crank, was just barely able to reproduce the nursery rhyme the inventor shouted into its mouthpiece (Figure V.3). Neither the business office nor the amusement center could have used a machine that was able to record less than two minutes worth of material and did so poorly. In the early 1880s Edison told his assistant Samuel Insull that the phonograph did not have "any commercial value."[1] Meanwhile, Alexander Graham Bell, Charles S. Taintner, and others

Figure V.3. Edison's earliest model of the phonograph (1877) consisted of three main parts: a mouthpiece (A) for recording the sound; a hand-cranked cylinder (B) covered with tinfoil on which the sound was recorded; a playback device (D) for replaying the sound recorded on the foil. This phonograph contained the bare essentials of a sound-reproducing system. Source: Edward W. Byrn, *The progress of invention in the nineteenth century* (New York, 1900), p. 274. Courtesy of Russell & Russell.

were laboring to improve the phonograph. Wax recording cylinders were introduced as well as a better holder for the recording stylus and a constant–speed electric motor. A dependable talking machine was finally made available to the general public in the 1890s.

The story of the unfinished state of the Edison phonograph could be repeated for many famous technological innovations: The cameras of the 1840s called for exposure times of ten to ninety seconds; the cumbersome and slow typewriters of the mid-nineteenth century were scarcely an improvement over writing with a pen; the first commercial internal combustion engine, the vertical Otto and Lan-gen engine of 1866, stood seven-foot tall and delivered three horse-power; the Wright brothers' first powered airplane stayed aloft only fifty-seven seconds; the television receivers of the 1920s displayed small images (1.5 by 2 inches) that were blurred and flickered badly;

and the first electronic computer occupied eighteen hundred square feet of floor space and weighed thirty tons. At first glance none of these appeared to be likely prospects for the basis of a new industry, yet all did so.

The selection of novelty involves risk and uncertainty. It rests upon an act of faith and the judgment that an invention will prove useful to some segment of the public and that it can be developed into a reliable device. We tend to hear about the times when that act was vindicated, when an Edison or Ford brushed aside doubts and criticisms to bring us the electric light bulb or the Model T. There are a number of occasions, however, when the public rejected a new product or when technological obstacles to a product's development could not be overcome. Limiting the sample to recent years and to automobile transportation, we find the combined airplane/automobile (late 1930s); the gas turbine-powered car, truck, and bus (1950s); the rotary internal combustion (Wankel) engine (1970s); and the alternative automotive power plants – modified steam and electrical – proposed as the solution to the energy crisis (1970s). All of these novelties received serious consideration by at least one of the major manufacturers; none of them has found a place in today's automobile.

Questions of crudity and utility aside, there are still many barriers to be surmounted before an invention becomes part of a people's economic and cultural life. In modern societies, capital, labor, and natural resources must be assembled, the working model converted into an acceptable consumer product, and the item fabricated so that it can be sold at a profit. Any one of these barriers may entail formidable problems.

The full account of the transformation of an invention into a finished technological and commercial product need not detain us, for as always our concerns are more broadly theoretical. Therefore, we shall turn to a general examination of some of the major factors influencing selection and to a discussion of the significance of alternative choices.

Economic Constraints

According to economic determinists, technological change is primarily a matter of demand: The market exerts a pull powerful enough to stimulate the inventor along certain lines of inquiry and then transform the invention into a finished commercial product. This explanation appears to be a plausible one, especially when it is applied to technological change in a modern capitalistic economy.

If there is a profit to be made on a given improvement or invention, some entrepreneur is bound to see its potential and produce it for the marketplace. In this deterministic view, the selection process is governed largely by economic forces.

Doubts arise about the cogency of the market-pull explanation as soon as we ask why certain novelties appear when they do. Are we to assume that a potential market for them never previously existed? Economic historian Nathan Rosenberg has noted that there are a host of deeply felt needs in the world at any time, needs that create potential markets, and yet only a very small number of these recurrent demands are ever fulfilled. Can we honestly claim, he asks, that there was never a potential market for modern high-yield grains, oral contraceptives, or heart pacemakers prior to the time they first appeared?

Market pull is certainly a mysterious force if it can exist in a potential form for decades, even centuries, and then suddenly bring forth a new product or device that "everyone always wanted." The power of the marketplace alone can no more account for the operation of the selection process than it could explain the emergence of novelty. In both instances it plays a role but by no means is it the main actor in the drama.

As we explore the selection of several important but diverse novelties – the waterwheel, steam engine, mechanical reaper, and supersonic transport – we will find that economic forces interacted with technological, social, and cultural factors in determining their selection. The selection of none of these novelties was controlled solely by economic demands nor was precisely the same collection of factors at work in each as the variant artifact was chosen and placed in a wider cultural context.

The Waterwheel and the Steam Engine

If a comparative study were made of the technologies of the great world civilizations ca. A.D. 1200, a single feature of Western technology would set it apart from Islamic, Byzantine, Indian, or Chinese technology. In thirteenth-century Europe, technology was marked by its extensive reliance upon waterpower. The vertical waterwheel, then dominating medieval European technology, had originated in the eastern Mediterranean region 150 to 100 B.C. and initially was used to power mills in which grain was ground into flour (Figure V.4).

Grinding flour by hand is a long and laborious task. In modern-day India a woman preparing enough flour for a single meal will

Figure V.4. Roman water mill of the fifth century A.D. (shown in partial cross—section). The direction and speed of the rotary power supplied by the vertical waterwheel were altered by gears and used to turn the horizontal millstone located above the wooden platform. Grain to be ground was poured into the hopper above the stones and flour was collected at the periphery of the stones. Source: Terry S. Reynolds, *Stronger than a hundred men* (Baltimore, 1983), p. 39.

spend two hours grinding grain in a hand mill similar to the one used in antiquity. Mills with larger and heavier grindstones can process grain more efficiently but require the expenditure of considerably more power, usually provided by donkeys or horses. Both methods require constant attention: Donkeys and horses needed to be fed and supervised, as did the women or slaves.

The water mill would seem to be an attractive alternative to the hand or animal mills. Although the building of a water mill required a higher initial investment, once installed there were few additional costs. Water flowed freely, except in droughts, and the water mill did not demand regular attention.

Despite these advantages water power was little used after it made its initial appearance. Not until the fifth or sixth century A.D. did the waterwheel come into its own. More than five hundred years elapsed between its invention and its widespread use in late Roman times. Why did a technological advance that was so useful remain dormant for so long?

Terry S. Reynolds, who has studied the development of early waterpower technology, offers several reasons for its slow diffusion. First, in antiquity the state of technical knowledge did not permit the construction of reliable and efficient wheels. Those built by Roman engineers were often not designed or constructed to make good use of the energy of the moving water.

Second, the Greeks and the Romans were restricted by their attitudes toward nature, labor, and technology. These were people who believed that nature, ruled by a panoply of gods, was sacred and not a realm open to casual human intervention and exploitation; the diversion of rivers and streams for waterpower might be interpreted as interference with the natural order. In addition, members of the wealthier and educated classes held manual labor in contempt and were reluctant to embrace technological innovations as a solution to their problems.

Third, economic reasons, according to Reynolds, might have played a crucial role. There was no tradition in antiquity for investment in technological improvements. Landholders were likely to hoard their money instead of putting it into a speculative venture involving an untried technology. Rather than jeopardize their accumulated wealth in such an investment, people utilized available labor, which was plentiful and cheap. Hand and animal mills were less expensive than waterpowered ones to construct and they had one advantage – in slack times horses, donkeys, or slaves could be readily sold and the money invested in them recovered. The argument that an abundant labor supply acted as a deterrent to the diffusion of waterpower is reinforced by evidence that in the fourth century A.D. water mills were recommended to Roman estate owners as a substitute for those run by human labor, because that labor was then becoming scarce and expensive.

At the time of the collapse of the Roman Empire, waterwheels were employed in a few places in Italy and southern France. In the Middle Ages the technology spread from these locales to the rest of Europe in a spectacular fashion. It had taken well over five hundred years for waterpower to establish a toehold in the classical world, yet in the seven centuries that followed the demise of the Roman Empire in A.D. 476 it was used on streams from Spain to Sweden, from England to Russia. What is most impressive is that these waterwheels were not confined to a few isolated sites scattered across the map. They are to be counted in the thousands, even the tens of thousands. The energy available in most large European streams was put to use in the Middle Ages.

Medieval millwrights and hydraulic engineers improved on the design of waterwheels and of the power dams and canals needed to bring water to them. Although this was an impressive development in itself, of still greater significance were the new uses found for waterpower in medieval industry. The Romans used waterwheels to grind their flour and lift water for irrigation purposes. By contrast, there were few aspects of medieval life that were untouched by waterpower technology. Wood was sawn, drilled, and turned by waterpowered tools; grains were ground and olives pressed in water mills; the tanning of leather, the making of paper, and the finishing of cloth employed waterpowered equipment; and mining and metallurgy depended on hammers, lifts, pumps, and bellows driven by waterpower.

The impact of waterpower technology on medieval society and economy was so profound that some modern historians claim that it was one of the major features of the era. The water mill, along with the windmill and more efficient harnesses for horses, are said to have constituted a power revolution that set medieval civilization apart from all earlier ones. For the first time in human history, a great civilization was built upon nonhuman power. Slaves did not bear the burden of medieval economic and cultural life because new power sources were developed to take the place of slave labor.

The widespread, diverse, and intensive use of new power technologies in the Middle Ages has led some economic historians to assert that an industrial revolution of sorts occurred in medieval Europe. Although this assertion may be rejected as an overstatement, we can still agree that in the Middle Ages the first decisive steps were taken. Medieval waterpower technology in large part laid the foundations for late eighteenth-century industrialization.

The question before us is not to uncover the exact relationship between medieval and modern economy and technology. We want to know why waterpower was so thoroughly incorporated into medieval culture when its selection was so steadfastly opposed in Greco-Roman times. The answer cannot be found in the progress of waterpower technology. The improvements made in it by medieval technologists were not crucial to its wide acceptance. The unimproved ancient waterwheel could have been used to transform the life and economy of Greece and Rome.

If the medieval period was the age of waterpower, it was also the age of faith. Any attempt to account for the rapid diffusion of the waterwheel throughout Europe during the Middle Ages must include in its consideration the influence of Christian tenets and institutions. Of particular importance was the establishment and spread of Western monasticism. According to the sixth-century A.D. Benedictine rules governing early monasteries, religious houses were to be secluded places where the monks could work and pray undisturbed. The Christian belief in the dignity of manual labor, a belief not common in antiquity, was central to monastic life. The brethren were expected to do most if not all of the work necessary to provide themselves with food, shelter, and simple material comforts. Given these circumstances, water mills served admirably. The use of the waterwheel ensured that the monastery would be a self-sufficient community with no need to traffic with the outside world, and left the monks with more time for their devotions. The monasteries of western Europe made early and extensive use of waterpower, not only to grind flour but to make beer, iron, leather, and cloth, among other things. The spread of the waterwheel technology was hastened by the building of monasteries in regions where it was hitherto unknown or little utilized.

Monastic water mills also provided examples for lay landowners faced with labor shortages and in need of new revenues. The water that powered monastic mills for the pious could also be put to work for the feudal aristocracy. The supply of labor, so abundant in antiquity, began to shrink in the turbulent late Roman era. By the early Middle Ages, labor scarcity had become a serious problem. Because much of the demand was in the agricultural sector, water mills proved useful in replacing a diminishing and more costly labor force. For that reason economic factors must be given equal weight with religious ones in explaining the growth of waterpower technology.

Medieval water mills, especially in some of their more elaborate manifestations, required the investment of a considerable amount of money. The feudal aristocracy was willing to make the investment because waterpower reduced labor costs and because mills could be made into a source of additional profits. Exercising their rights over the serfs who worked the land, the feudal lords compelled them to use the manorial grain mill and pay in kind for its use.

At the end of the Middle Ages, another wealthy social group, merchants living in urban trading centers, were also in a position to invest in waterpower. Because this new group could not rely on a manorial monopoly for income, they used the waterwheel in pursuit of their commercial interests, thus bringing it to new industrial and manufacturing applications.

The diffusion and uses of the waterwheel cannot be explained on economic grounds alone. Both economic and cultural factors must be considered together as we try to understand how an innovation that was ignored by the ancients became the focus of a power revolution that drastically altered life in medieval and early modern Europe.

The reign of the waterwheel was ended by the selection of the stationary steam engine, but the emergence of this new source of power was not as abrupt as we are sometime led to believe. For many decades the two coexisted in European and American industries. The eventual triumph of steam occurred more than a century after the appearance of Newcomen's engine.

The 250-year period between 1500 and 1750 marks the high point of the industrial use of the waterwheel. Building on the achievements of the Middle Ages, the industrial economy of the sixteenth, seventeenth, and eighteenth centuries advanced rapidly with waterpower driving it forward. In western Europe and Britain the number of water mills multiplied, waterwheels were designed to deliver more power, and the industrial applications of waterpower grew. All of this was accomplished with wooden waterwheels built by millwrights who relied on the accumulated knowledge of centuries of wheel and dam construction.

After 1750, under the dual impact of heightened industrial growth and competition by the steam engine, the waterwheel was transformed into a much more efficient and modern power source. Systematic experimentation with various wheel designs, the theoretical analysis of the waterwheel in terms of the hydraulic principles on which it operated, and the substitution of iron for wood in the construction of the wheels led to a more sophisticated waterpower technology. The improved waterwheels could easily compete with the typical early nineteenth-century steam engine.

The first working Newcomen engine was erected in 1712 and used for a very specific purpose – pumping water from mine pits. Mining was one industry that could not accommodate itself to the limitations of waterpower. Textile factories could be sited next to good waterpower sources but not coal or tin mines. At best canals might be dug to carry water to the mine's waterwheel or power generated by a close-by stream could be transmitted by a system of linked rods (called *Stagenkunst*) to operate pumps at the mine. Neither method was completely satisfactory, consequently, the Newcomen engine, often burning the fuel mined on the spot, first found a place in mining regions.

Many of the industrial applications of waterpower relied upon the smooth, constant, rotary motion of the waterwheel. The motion generated by a Newcomen engine was a reciprocating, or back and forth one, well suited to operating pumps but not factory machinery. The problem then was to obtain rotary motion from a steam engine. The immediate and ingenious solution was to use a steam engine to lift water for the continuous operation of a waterwheel. In this way, two disparate sources of power were joined to create the rotary motion needed to turn the machinery in the factories or the grindstones in flour mills. Steam engine–waterwheel combinations were quite popular in late eighteenth-century Britain, especially in the textile industry.

As a result of the inventive work of James Watt between 1780 and 1800, smooth rotary motion could now be obtained from a dependable and efficient steam engine. These developments eventually led to the displacement of the vertical waterwheel from its dominant role as a supplier of industrial power.

Centuries of improvement, however, could not overcome the most serious liabilities of waterpower. Droughts, floods, and ice all interfered with the operation of the waterwheel. Geographical location also had an impact, because waterpower was limited to larger, fast-moving streams. A final liability involved power production. Improvements in design had maximized the possible output, yet more power was needed for the new machines. The unit power output for waterwheels was roughly equivalent to that of most steam engines during the first half of the nineteenth century. As the size of the mills grew and more horsepower was needed to run the machinery, the steam engine was able to meet the new challenge while the waterwheel lagged behind.

These drawbacks were of crucial importance yet they could become irrelevant under special circumstances. In France and the United States, where there were many new waterpower sites that could be easily developed, the steam engine was adopted later than in Britain, a country with few such sites. Generally speaking waterwheels were cheaper to erect than steam engines, they required little maintenance, were not as prone to failure as early steam engines, and were a familiar and well-tested source of power. For these reasons the waterwheel survived and managed to offer stiff competition to the Newcomen and the Watt machines for over a century.

The waterwheel, which had provided much of the energy needed for the industrialization of Europe and America, could not meet the power requirements of heavy industry after 1850. Ironically, the very success of the Industrial Revolution created power demands that could only be met by the steam engine.

There is a lesson to be learned from the resistance the steam engine encountered as it sought to replace the waterwheel. A modern observer caught up in the ideology of technological progress might well assume that the steam engine should have rapidly displaced the waterwheel. However, as we have seen, the end of waterpower was not forestalled by conservative factory owners who kept it alive beyond its time. On the contrary, the waterwheel was retained for decades in the nineteenth century because it made good economic and technological sense to continue using this power source, which had appeared before the birth of Christ and had been extensively used since the Middle Ages.

The Mechanical Reaper

In nineteenth-century America the creation, selection, and development of a technological novelty was often carried out by a single figure – the inventor–entrepreneur. People like Robert Fulton, Samuel F. B. Morse, Cyrus H. McCormick, and Thomas A. Edison were as active in their workshops as in business circles where they raised the capital needed for the manufacture of their inventions. Not satisfied to sit idly by while others weighed the merits of their inventive labors, these men used their talents and energy to hasten the process of adoption of their innovations by society. A case in point is Cyrus H. McCormick (1809–84), the inventor of the first generally successful and widely used mechanical reaper and also one of the founders of the mechanical harvesting industry.

Until the middle of the nineteenth century, most harvesting was done by persons who walked into the grain swinging their long-handled scythes, slicing through the stalks that stood before them. Each swing of the scythe cut a semicircular swath of grain. The English were the first to attempt the mechanization of reaping and tried to duplicate the harvesters' swinging action with machines equipped with rotating scythe blades (1790s). When those failed, they built mechanical reapers with scissor-action cutting blades. The latter had a limited success and, in the 1830s, were soon supplanted by McCormick's machine whose cutter employed a serrated or notched blade in a sawing motion.

Having first established the basic design for a workable reaper in 1831, McCormick began improving his machine and laying plans for its production (Figure V.5). Recognizing that the reaper market in the eastern United States was limited by the number of small-

Figure V.5. The McCormick reaper of 1840. The reaper was drawn through the crop to be harvested with horses treading on the stubble portion of the fields. The forward motion of the machine activated the rotation of the large reel, which pushed the stalks between the projecting teeth of the cutting mechanism (not shown here). Source: Michael Partridge, *Farm tools through the ages* (Reading, 1973), p. 129, by permission of the University of Reading, Institute of Agricultural History and Museum of English Rural Life.

sized farms located there, he left his native Virginia for the Midwest and the large wheat fields of the prairie states. The entire McCormick manufacturing operation, which had begun slowly in the East, was transferred to Chicago in 1847 once the inventor found the financial backing he needed to open a shop in which reapers could be produced in quantity. As his business prospered, McCormick paid less attention to the technical side of it, leaving such matters to the mechanics and engineers he hired, and concentrated on legal battles over patents and the promotion of reaper sales.

McCormick was a pioneer in the creation of new business techniques. He made good use of field trials that pitted his reapers against those made by rivals, searched for novel ways to advertise his product, and offered special financial incentives to purchasers of his machines. The McCormick Harvesting Company soon became a dominant force in the agricultural equipment industry. In 1902 it merged with Deering Harvester Company to form the industrial giant International Harvester, since renamed Navistar International.

The often-told story of Cyrus McCormick, a success story that rightly stresses his reaper's impact on grain production, obscures some of the problems the entrepreneur faced during the early years of reaper manufacture and marketing. Given that mechanical reaping replaced inefficient, labor-intensive harvesting with the scythe, one wonders why McCormick's machines were not bought in large numbers by farmers before the 1850s. After all, the mechanical reaper could cut ten to fifteen acres of wheat in one day compared to the hand-harvester's one to two acres. The McCormick reaper was patented in 1834 and production began seven years later. Yet few reapers were sold between 1841 and 1855.

In large part this hiatus can be explained by the troubles the inventor met as he developed his prototype into a fool-proof agricultural machine that would function over different terrains and could be maintained by the ordinary farmer. Economic historian Paul A. David has suggested that this was not the sole reason for the slow diffusion of the reaper. He argues that through the early 1850s it was more profitable for farmers with smaller holdings to use scythes to harvest grain than it was to purchase a mechanical reaper. According to his calculations a farm of 46.5 acres marked the threshold size in reaper use. Only those who worked more than the threshold acreage found it worthwhile to invest in mechanical reapers; the others continued to rely on human labor. After the mid-1850s labor costs rose, the size of grain farms increased, and the price of the reaper remained stable, at which point it became economical to use mechanical reaping devices.

If the history of the McCormick reaper depicts the activities and triumphs of a typical nineteenth-century American inventor—entrepreneur, it also points to his limitations. No matter how effective and dedicated McCormick was as a promoter of his invention, its ultimate acceptance depended upon the machine's technological and economic feasibility.

The reaper appeared on the scene as the cultivation of wheat was moving westward, spurred by the vast stretches of cheap, fertile land. These farms were large, relatively flat, and devoid of rocks making them well suited to mechanical harvesting techniques.

The introduction of the mechanical reaper also coincided with other technical improvements in agriculture and with the extension of the rail system. Railroads were crucial to the growth of agriculture in the Midwest and West. Surplus grain was shipped by rail to the populated urban centers, and manufactured goods and finished lumber were transported to the rural areas. From this perspective

McCormick's reaper can be understood as an important factor in the westward expansion of the nation and in the industrialization and urbanization of late nineteenth-century American society.

The Supersonic Transport (SST)

In the twentieth century the selection of technological innovation had become much more complex than it had been in the nineteenth century, as is revealed by the controversy surrounding the proposal that the United States build commercial aircraft capable of flying faster than the speed of sound. No strong-willed inventor-entrepreneur could have gained control of the unwieldy set of opposing institutions, individuals, and constituencies that clashed over the supersonic transport (SST).

The SST is not a typical example of how selection operates in the case of a modern technological development yet it encompasses most of the elements one finds scattered throughout a number of more representative samples. The major participants in the SST debate included the government, industry, and the public. Their disagreements over technological and economic issues were heightened by the different views they held regarding the importance of national prestige, long-term economic growth, the preservation of the environment, and the quality of life. The final outcome of this lengthy debate (1959–71) was the decision not to build the SST.

That the selection process includes the rejection, as well as the retention, of novelty is essential to our understanding of the nature of technological change. The deliberate rejection of novelty serves as a corrective to those who think that technology moves smoothly from one success to another. Making a selection from among competing technological possibilities necessitates that some of those possibilities must be excluded from the mainstream of made things.

In the late 1950s few challenged the prediction that American-built SSTs would soon be transporting passengers around the world. Ever since the days of the first flights at Kitty Hawk in 1903, aircraft speeds had increased at a steady incremental rate. Therefore, when the first serious consideration was given to SST development, people did not question the idea that by 1970 transport aircraft would be flying at speeds of Mach 2 or Mach 3. (Mach 1 equals 760 miles per hour at sea level.)

The American aircraft industry had flourished during World War II by developing and producing military planes under government contract. The Russian–American confrontation during the postwar years helped perpetuate the close collaboration between the military and aircraft manufacturers. This collaboration had established the precedent for government subsidization of the development of radically new and faster aircraft. Although the supersonic transport was to be a civilian plane, the aircraft industry assumed from the outset that the funds for its development would come from the federal treasury, because new airplane designs had received government funding in the past.

In the late 1950s, the Boeing, Douglas, and Lockheed aircraft companies began feasibility studies on the SST. Their expectation that the government would finance the undertaking was reinforced by rumors that Great Britain and the Soviet Union had already embarked upon their own state-supported SST programs. From among the potential governmental sponsors for the SST project, the Federal Aviation Agency (FAA) emerged as the body that would carry the responsibility for SST development.

In the closing months of the Eisenhower administration, it was generally understood that the FAA would oversee the work on a stainless steel–titanium aircraft with Mach 3 capabilities that would surpass anything being planned by rival nations. This technological marvel, as yet unfunded, was expected to begin carrying commercial passengers in eight or nine years.

In the new administration of John F. Kennedy, the FAA continued to promote the American SST. This job was made easier by the November 1962 announcement that the French and British governments had agreed to the joint development of the aluminum Mach 2.2 Concorde. Officials at the FAA predicted that if the government did not act quickly the nation would lose the potential international market of 210 to 250 commercial planes, fifty thousand aviation-related jobs, and its leadership role in world civil aviation. Despite these dire predictions, there were skeptics who were opposed to spending large amounts of federal money on what they saw as a purely commercial enterprise, and a doubtful one at that.

Then in July 1963 Pan American Airlines placed an order for six Concorde transports. This was a move calculated to force the United States government to make a financial commitment to an SST program. Shortly thereafter, President Kennedy declared that the federal government, Congress willing, would underwrite most of the development costs of an American-built SST. Of the one billion dollars needed, 75 percent would come from the federal treasury and the remainder from private industry. The FAA immediately issued specifications for SST engine and airframe competitions, announcing optimistically that commercial flights would begin before June 1970.

During the next five years, from 1963 to 1968, the SST program underwent a period of fragmentation. Hidden doubts and animosities appeared, bureaucrats defending their turfs squabbled with one another, public confidence waned, and technical problems persisted. The resolve to build the SST, and do it soon, was lost.

Troubles began with the manufacturers who were not satisfied with the cost distribution formula of 75 to 25 percent. They wanted, and eventually got, a funding arrangement of 90 to 10 percent. After the Kennedy assassination, the new Johnson administration decided to take a fresh and harder look at the entire SST project. In the process the FAA lost absolute control over the SST to a presidential advisory committee. In the advisory committee questions were immediately raised about the commercial soundness of the venture. Would customers pay higher SST fares when they might travel more economically on the jumbo jets that would soon be available? Economists were not certain they would.

The sonic boom, an explosive sound created by aircraft flying at supersonic speeds, also became a serious public relations problem at this time. The FAA sponsored a series of sonic boom tests over Oklahoma City and were told by 23 percent of the people who experienced them that they could never learn to live with the noise. Only the work of enthusiastic SST proponents in Washington, and the threat of foreign SSTs in American skies, kept the program alive. The foreign threat gained credibility in December 1968 when the Soviet Union's supersonic TU-144 made its maiden flight and in March 1969 when the Concorde did likewise.

Mel Horwitch, historian of the American SST conflict, has labeled the next stage in the debate (1968–71) quite simply: "Explosion." At this point, the SST became an issue of intense public concern. The heated debate finally ended with Congress voting to cut off all funds for the aircraft's further development. The source of the explosion was the entrance of well-organized public interest groups into the question of SST development. The informed criticism of these groups centered on two elements: the deleterious effects of sonic booms on the health and property of those living along the paths of SST travel, and the damage done to the upper atmosphere by supersonic flights in that region. The already slowed SST momentum was halted by the intrusion of this new element into the debate.

In 1968 President Nixon had assumed office, becoming the fourth president to deal with the SST issue. His initial response was a request for ninety-six million dollars for two SST prototypes. The president's action served to intensify wide public criticism of the SST. In April 1970, a broad-based coalition against the SST was established in the nation's capital. Anti-SST public sentiment was galvanized as respected scientists and government officials testified in Congress against the allocation of any federal funds for what appeared to be an economic and environmental disaster. Under this intense pressure, Congressional friends of the SST wavered and by a series of close votes, the last one in May 1971, the program to build a supersonic transport was canceled.

The proponents of the SST were surprised by the intensity and effectiveness of the public criticism of their pet program. They need not have been. The late 1960s and early 1970s were marked by both organized and spontaneous public protests over the war in Vietnam, civil rights issues, and the destruction of the environment. The SST symbolized big government acting in behalf of big business and unbridled technology without regard for the rights and well-being of ordinary citizens. In their disappointment, the SST supporters accused their opponents of being Luddites bent on destroying one of the best hopes of Western technology. This was not true; yet never in modern times had there been such a clear and concrete public challenge to the belief that technological change was progressive and inevitable.

Horwitch lists a number of reasons why the American supersonic transport was rebuffed. The SST, which was not fuel efficient, encountered fuel prices that were rising and an economy that was declining. When confronted with these problems, advocates were never able to offer good, strong economic reasons for the SST. The FAA, essentially a regulatory agency, was unprepared to do this job, and its managers were insensitive to the issues motivating protest groups at the time. The military made no defense claims for the construction of the SST, and the airlines, doubtful of the SST as a profitable commercial enterprise, were reluctant to embrace it wholeheartedly. If these were not enough to kill the program, there were also the technical flaws in the aircraft's design and the politically ominous prospect of the sonic booms riling the populace as the SST flew overhead.

In reviewing the many reasons for the SST failure, one might well ask why the friends of supersonic travel were so oblivious to them. The answer is twofold. First, they were enthusiastic believers in technological progress. We had always built faster aircraft in the past and given the chance we would do so again, especially if the government was footing the bill. Second, they had not considered that the public, through articulate, concerned activists, would enter what they supposed would be a restricted debate. According to the prevailing rules, elites in business, the military, and the government should handle these matters. The process of selecting technological innovations for further development was not normally thrown open to public discussion, even when the taxpayers' money and welfare were involved. If the SST debate did not radically alter that traditional selection process, it at least exposed its inequities and set the stage for possible public intervention in similar cases in the future.

The SST is also an excellent example of the way in which economic factors can become entangled with, and overwhelmed by, other forces in the selection process. Long before there existed organized opposition to the SST, doubts were expressed within the government and the airline industry regarding its profitability. Proponents of the SST countered by offering rosy economic projections, shifting the discussion to the progress made in the SST programs of rival countries, and using bureaucratic and political maneuvers to advance their project. When the subsidized aircraft manufacturers failed to show sufficient enthusiasm for supersonic travel, the FAA prodded them to issue positive evaluations of its worth. During the final phase of the debate, a distinguished group of American economists released an influential set of statements critical of the SST effort; however, the initial decision to develop the SST was not based on an objective study of its economic worthiness, and the final defeat was not the result of a preponderance of economic arguments advanced against it.

The SST was not a lunar lander or some new missile for America's defense arsenal; it was to be part of the business of transporting paying passengers to their destinations around the world. That so blatant a profit-making endeavor could be advanced so far on such a slim basis of economic promise raises questions about how many less-commercial technological projects have been selected for development on yet more shaky economic grounds.

Military Necessity

Commercial viability understandably recedes in importance when we turn to military technology. A concern for costs and adequate returns from an investment is overridden by urgent military necessity during wartime or the need to preserve a nation's security when it is at peace.

In the modern era, military imperatives have affected the selection of key technological innovations that eventually found a place in the civilian world. Therefore, the military and civilian aspects of modern industry are intimately connected. Historians generally agree that military technology has evolved rapidly in recent decades, but they disagree about its long-term impact on economic growth, industrialism, and the development of other technologies. For some historians war is a force that stifles civilian industrial growth. For others it is an essential element in the establishment of industrial capitalism. The latter argue that nineteenth-century industry modeled itself on the military using factories instead of barracks, laborers instead of soldiers, and corporate planning and strategy instead of military planning and strategy.

Early proof of the association of the military with advanced technology is found in the Renaissance machine books, which portrayed fantastic engines of war. Somewhat later military demand for large quantities of standardized clothing, foodstuffs, and weaponry did prefigure, in a sense, the creation of mass markets fed by mass production. The metal industry was often closely tied to weapon manufacturing, as was the explosives industry. And wartime profits did launch many new industrial enterprises. Nevertheless, it cannot be proved that military necessity of and by itself was responsible for the origin of modern industry.

In my study of the pressures exerted by the military on the selection process, I have chosen to skirt the controversies associated with the broader issue of warfare and industrialism, and concentrate instead on how two specific innovations were first adopted by the military and later by civil society – the motor truck and nuclear energy.

The Motor Truck

The early years of the twentieth century saw the beginnings of the age of automobility in the United States. Automobile historian James J. Flink has noted three important dates in that period: 1905, when the annual New York Automobile Show became the nation's largest industrial exhibit; 1907, when the car was perceived by Americans to be a necessity; and 1910, when 485,377 automobiles were registered in America making it the world leader in car ownership.

Motor trucks, in America and elsewhere, lagged behind automobiles. In 1910 when automobile registration was approaching the half-million mark, truck registration stood at 10,123. In the first decade of the gasoline age, cars far outstripped motorized vehicles expressly designed for goods transport.

Experiments with motorized truck transport were carried on simultaneously with work on the early automobile. Trucks of varying sizes and body styles, with gasoline, steam, and electric power plants, were built and put to commercial uses in the late 1890s. As early as 1900 a distinguished professor of engineering at Cornell University, in a comparative study of the truck and the horse-drawn wagon, concluded that operating costs for the former were 25 to 40 percent lower. In 1904 the American Express Company released the results of privately conducted tests that showed that motor-truck delivery was superior to horse-and-wagon hauling, and in 1909 an article in *Scientific American* announced that the truck had become an economic necessity for businessmen. Although motor-truck sales rose steadily, overall they remained relatively small; business and industry continued to rely on the traditional and proven horse-drawn wagon. This situation obtained until the truck went to war, first against Pancho Villa and his Mexican rebel army (1916) and then against the Central Powers (Germany, Austria-Hungary, Turkey) in World War I.

Military planners had not yet abandoned the horse, or mule, and wagon when World War I opened. Still, England and France immediately asked for shipments of cars, trucks, and ambulances, a request that stimulated American truck manufacturing. In the meantime President Wilson ordered Brigadier General John J. Pershing into Mexico to capture Pancho Villa after his attack on American territory. For the pursuit, and its supply lines, Pershing requested 70 motorized truck trains, with 27 trucks to a train. His order for 1890 trucks came at a time when motor vehicles assigned to all Army commands totaled 1000. By the end of Pershing's fruitless Mexican campaign, the Army had 2700 trucks and the U.S. was preparing to enter the war in Europe. American manufacturers had already sent 40,000 trucks and ambulances to the Allies when Pershing, assuming command of the American Expeditionary Force in 1917, called for at least 50,000 additional trucks to transport troops and supplies from railheads to the front lines.

The battlefields of World War I served as proving grounds for truck engine, transmission, and body designs; military requirements helped to bring uniformity and standardization to truck manufacturing; wartime uses of the truck demonstrated its dependability and adaptability; and government contracts financed the expansion of truck production facilities. An editorial writer in a 1918 issue of *Horseless Age*, the prime journal of American motoring, declared: "This war has advertized the motor truck to the world more than anything else ever could."[2] Later, a historian writing about the beginnings of the trucking industry concluded: "World War I was the cradle in which the motor truck was nurtured."[3]

In 1914 America's truck makers produced 24,900 units; four

years later their production figure was 227,250 units. Armistice found truck manufacturers geared for capacity wartime production. Peace meant an end to lucrative government contracts and the potential that the market at home and abroad would be filled with war-surplus motor vehicles no longer needed in a smaller peacetime Army.

Despite enormous problems in the postwar years, American truck production thereafter never fell below the wartime record set in 1918, except for the economic depression year of 1921. National defense was invoked in the years immediately following the war to gain federal money for nationwide highway construction programs. General Pershing, in testimony before Congress (1921), argued that the country's defense depended upon a network of good roads, and a year later the Army produced the "Pershing map" showing the main arteries that were vital to American security in wartime. These so-called national defense roads conveniently coincided with the principal routes already designated by the states for incorporation in a federal-aid highway system. The War Department assumed that highways that met the country's industrial and commercial requirements would also serve its military needs. The Army's endorsement of postwar federal highway plans aided the nascent motor-truck industry's attempt to become a major transporter of goods over long distances, in direct competition with the railroad.

Nuclear Power

Unlike the motor truck that was in existence before the Army made it popular in World War I, the nuclear power reactor was a direct outgrowth of the military uses of nuclear energy. America would not have a nuclear power industry today had the atomic bomb not been developed during World War II. Indeed, military necessity has been enormously influential in shaping international nuclear power production for nearly fifty years.

Hitler's rise to power in Germany during the 1930s forced some European and American scientists to conclude that the German dictator might be the first to manufacture an atomic bomb for military purposes. Accordingly, these men urged their governments to initiate programs whose aim would be to move atomic energy from an experimental stage to its practical application in a full-fledged weapon.

One of the key figures in the promotion of the military aspects of nuclear energy was the Hungarian physicist Leo Szilard who had fled Hitler's Germany in 1933. Szilard realized that a self-sustaining nuclear chain reaction was capable of liberating vast amounts of energy for large scale military and industrial uses. Therefore, in October 1939 he drafted a letter to President Franklin D. Roosevelt, to be signed by Albert Einstein, informing the American leader that recent developments in nuclear physics made it possible to construct a new type of extremely powerful bomb.

Within three years of the Szilard–Einstein letter, the U.S. Army was given the responsibility for directing the Manhattan Project to create an atomic bomb. The project, employing more than a hundred thousand men and women, spent over two billion dollars and eventually developed the bombs dropped on Hiroshima and Nagasaki (in August 1945). Any hope of cheap, abundant nuclear energy for civilian use vanished as scientists and engineers under the leadership of General Leslie R. Groves focused all efforts on the making of the first atomic bomb.

The scientific knowledge and technical expertise that enabled the Manhattan Project to accomplish its military mission could, to an extent, be transferred to peacetime nuclear power production. The isolation of the uranium U-235 isotope, the one best suited for fission, presented a formidable technical barrier that was overcome by building very large and costly industrial plants for isotope separation. Uranium-235, originally employed as an explosive in the Hiroshima bomb, also served as a reactor fuel in postwar power plants.

The world's first nuclear reactor, the graphite and uranium pile erected by Enrico Fermi at the University of Chicago in 1942, was built under the sponsorship of the Manhattan Project. Larger and more sophisticated reactors were constructed later by the project for experimental purposes and for the production of another bomb material, plutonium. Even though power generation was not the sole reason for the existence of these reactors, the knowledge gained in designing, building, and operating them proved useful to the nuclear power industry.

In the years following the dropping of the atomic bomb on Hiroshima, popular-science writers wrote longingly about an atomic paradise free from disease, poverty, and worry. Journalists and government spokesmen promised that all of this could be achieved if humanity abandoned the military uses of nuclear energy and embraced the peaceful atom.

Contrary to popular dreams, the postwar political situation dictated that the military atom would continue to reign supreme, a situation paralleled by the beginning of the Cold War. East–West tensions alone, however, cannot be blamed for the postponement of the atomic paradise. Reactors had not yet advanced to the point where they could safely and effectively produce electrical power. Furthermore, there was some question as to whether uranium was available in sufficient quantity to fuel a number of large reactors.

While power reactors languished in their experimental stage, nuclear weapons technology raced forward in the United States and the USSR. Faced with an escalating arms race, President Eisenhower suggested that the two superpowers, in the interest of peace, share their nuclear technology and materials with the rest of the world. The plan was never adopted, but the Atoms for Peace program of late 1953 that grew out of it was an American propaganda triumph.

An important part of Atoms for Peace was the American promise to help other nations, especially underdeveloped countries, to build reactors for power production. The truth of the matter was that in 1953 the United States had no power reactors of its own, hence no models for export. Meanwhile, both the Russians and the British were said to be close to generating electricity with reactors they had designed. If Atoms for Peace was to remain viable, and if the United States was to maintain its leadership in nuclear technology, then it was necessary to develop a suitable American power reactor quickly. The solution to this problem came, as it only could come, from an ongoing military project. In this instance, it was one under the sponsorship of the U. S. Navy.

Whereas the Army was spending billions of dollars producing nuclear weapons, the Navy had not been given an opportunity to enter the atomic age. Once peace was declared, the Navy, determined not to be bested by rival Army, turned its attention to nuclear energy as a means of propelling surface and underwater craft. This required the careful control of nuclear reactions so that the heat liberated could be used to convert water into steam, which could then be fed into the ship's conventional propulsion turbines. Given its aims, the Navy was understandably far more concerned with reactor technology than it was with the design of nuclear weapons.

In the early postwar years Admiral Hyman G. Rickover was not the only naval officer intrigued by the idea of a nuclear navy, but he, more than any other individual, can be credited with making that idea a reality. Moreover, the engineering decisions Rickover made as he developed his nuclear propulsion systems, and the auspices under which he worked, were to have an impact on the international nuclear power community.

Rickover, an engineering officer who, in 1947, rose to head the Navy's nuclear submarine program, was called upon to make a critical choice in 1950. He was forced to select the type of reactor to be used on the world's first nuclear submarine, the USS *Nautilus*. This decision was to prove extremely important to the later history of nuclear power production.

In a power reactor, apart from safety considerations, the main technical problems are the control and moderation of the nuclear reaction, the maintenance of an acceptable temperature at the reactor's core, and the transfer of heat energy away from the core so that it can do useful work elsewhere. In 1950 one could remove the heat from a reactor core by using ordinary (or light) water, heavy water (deuterium oxide), a liquid metal, or a gas in a variety of heat exchange systems. Each of the systems had its advantages and disadvantages and none of them had been tested extensively. After all, reactor technology had been in existence for only eight years.

Rickover, who promised to have a nuclear submarine in the water by January 1955, could not afford to make a mistake in choosing its power reactor. After carefully investigating the available options, he selected a reactor using ordinary water as coolant and moderator. Responsibility for its construction was given to the Westinghouse Corporation of Pittsburgh, Pennsylvania. The choice of a so-called light-water reactor was a conservative one, made by an engineer who knew that more technical data were available on water than on some of the more exotic coolants and that the technology for water transfer already existed for steam boilers, turbines, and the like. Rickover's decision led to a spectacular success. He was able to put his nuclear submarine into operation on 17 January 1955. As it sailed around the globe the USS *Nautilus* broke all previous records for underwater travel.

Submarines were but one part of an all-nuclear navy. The nuclear propulsion of large surface vessels, specifically aircraft carriers, also became a Navy priority. Impressed by the progress Rickover was making on his nuclear submarine, the Navy chose a light-water reactor for its proposed nuclear carrier (Figure V.6). The carrier's reactor was to be built and tested as a shore-based prototype prior to its being prepared for shipboard installation.

Before the carrier project had progressed substantially, it was canceled by the Eisenhower administration in an economy move. By a series of adroit maneuvers, however, Rickover revived the carrier reactor in a new form. Now it was to be an entirely civilian enterprise, the first American reactor to feed power into the national electrical grid. With the backing of the Atomic Energy Commission, the agency formed in 1946 and put in charge of all American atomic energy activities, and with the blessing of the president who needed

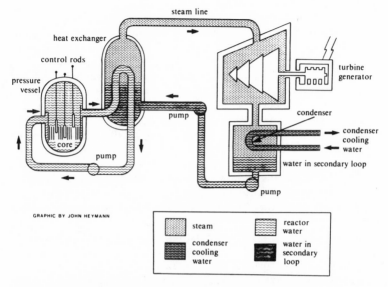

Figure V.6. Schematic diagram of a light-water reactor. In this model, water under pressure is pumped along a closed (primary) loop carrying the heat generated by the core to the heat exchange. In the heat exchange water in the secondary loop is converted into steam, which is used to run the turbine generator. Steam leaves the turbine generator to be condensed and pumped back to the heat exchange. As depicted here the steam is used to generate electricity; in a submarine the steam is used to propel the craft. Source: Stephen Hilgartner, Richard C. Bell, and Rory O'Connor, *Nukespeak* (San Francisco, 1982), p. 114.

such a reactor for his Atoms for Peace proposal, Rickover embarked on another successful crash program. This one laid the foundations for the American nuclear power industry.

Working with Westinghouse, as the fabricator of the reactor, and with Duquesne Light Company of Pittsburgh, Rickover drew up plans for America's first commercial nuclear power reactor. It was to be built on the Ohio River at Shippingport, Pennsylvania. Ground-breaking ceremonies took place September 1954 and by Christmas 1957 the reactor was generating sixty megawatts of electricity using a modified version of the nuclear-carrier reactor.

The Shippingport generating station was of decisive importance in shaping the nuclear power industry for decades to come. Its reactor served as prototype for later ones built and used in the United States and those sent abroad by American companies. Shippingport utilized a light-water reactor and so did a majority of plants thereafter. Of the nearly 350 reactors operating in the world, about 70 percent of them are of the light-water type.

Rickover initially chose water as a coolant–moderator because it met his immediate needs for submarines and not because a light-water reactor was the best type to supply power for electrical utility companies. Indeed, light-water reactors are one of the least efficient consumers of uranium fuel. The choice of a reactor for Shippingport was strongly influenced by the United States's determination to get a show-piece nuclear power plant in as short a time as possible. The Navy happened to have a working reactor for the propulsion of its vessels and that one was hastily converted for a far different use.

To understand how different was the use it is necessary to question the economics of the light-water reactor. Such a query, of course, has little meaning within a military context. The builders of atomic bombs and naval propulsion units, although expected to stay within their ample budgets, need not worry excessively about economic matters. Military necessity and national security have first priority. Shippingport, on the other hand, was part of the competitive business of producing electricity for the marketplace.

How did America's first nuclear power plant fare when compared with its competitors? In the late 1950s, when a coal-fired steam plant could produce electricity at a cost of six mills per kilowatt of generating capacity, at Shippingport the cost was sixty-four mills per kilowatt. Because the technology was new, costs were expected to drop as it was refined, and they did. Yet after thirty years coal-fired plants can still produce power at slightly lower costs than nuclear ones. The discrepancy is no longer tenfold but neither are we on the verge of producing electricity that is too cheap to meter, as the early proponents of nuclear power promised. Some critics have alleged that the Shippingport example saddled the industry with a light-water reactor, a model doomed to be forever uneconomical. That may or may not be the case. What is certain is that when the original decision was made, the economics of reactor operation was one of the last aspects to be considered.

Nuclear energy's long military associations exerted other influences on it as it moved into the civilian sector of the economy. Research and development of nuclear energy since the advent of the Manhattan Project had been generously subsidized by the federal government. Large American corporations have become accustomed to a situation in which the government took the financial risks when nuclear energy is involved.

The Shippingport reactor, ostensibly part of a commercial complex, was owned by the Atomic Energy Commission. The electric generating facilities were built at Duquesne's expense. The company invested its own money hoping to benefit from the publicity surrounding this pioneering effort and expecting the commission to underwrite operating costs if the venture turned out to be an unprofitable one. Duquesne invested in the conventional technology of Shippingport, the least expensive and safest part of the project.

Reactor research and development was very costly, and private industry has balked at subsequent government attempts to induce them into financing new reactors for larger generating plants. The electrical industry cannot be blamed for its reluctance to invest in nuclear technology. Existing coal-fired plants were based on a well-known and thoroughly reliable technology, and there was no shortage of fuel looming in the near future. The nuclear power industry was not born out of a desperate need of the electricity producers to find an alternative to traditional energy sources.

Although it is the American experience that has been recounted here, the interaction between the military and civilian applications of nuclear energy is not limited to the United States. As early as 1940, a British report on atomic energy declared that a close relationship existed between the exploitation of nuclear energy for military purposes and for power production in peacetime. "The development of one will," it concluded, "have a considerable effect on the development of the other."[4]

At the time that the Shippingport plant was being planned, the Soviet Union, Great Britain, France, and Canada were developing reactors for electrical power production. These nations had different social institutions and political traditions, yet their power reactors were all closely tied to military programs. The Soviet reactor was adapted from a naval propulsion unit, the British and French models were based on reactors originally built to produce plutonium for bombs, and the Canadian reactor had been indirectly and heavily subsidized by the American government through its purchase of Canadian plutonium for the making of weapons.

Without the pressure of military necessity, and its accompanying largess, there would be no nuclear power industry today. It is difficult to imagine any set of circumstances, short of war or a disastrous energy shortage, that would have forced the American government in 1941 to commit its material resources, manpower, talent, and money to transform the physicists' chain-reaction experiments into a working bomb or reactor. We who live on the other side of the nuclear revolution forget that before it gained its postwar glamour, nuclear physics was a rather esoteric field of study. It had its popularizers and promoters, but they could never have persuaded private industry or the federal government to spend two billion dollars on nuclear energy research over a four-year period.

Even with the war and government money to spur them on, those who worked for the Manhattan Project were not always certain that its goals were attainable. In a peacetime setting, with limited funds and personnel, technical problems that temporarily disheartened participants in the project would have loomed as insurmountable barriers.

In the second half of the twentieth century the distinction between the economic and military factors affecting the selection of technological novelties has been blurred. Earlier, the military was in ascendancy as a selecting agency only in wartime and during the preparation for war. At other times it made few demands on technology, as exemplified in our discussion of the motor truck.

After World War II came the Cold War, the arms race, the space race, and the belief that national security required an ever higher technological level of military preparedness. In the warfare state that now exists in the more powerful industrialized nations, innovations are constantly examined for their military potentiality and major industries are devoted exclusively to serving military markets.

Many of the most exciting new technologies of the late twentieth century bear the stamp of their military origins. They include jet-propelled airplanes, spacecraft, radar, computers, numerically controlled machine tools, and miniaturized electronics.

The extraordinary role played by the military in determining technological choices makes our age unique in the history of technology. Never before have so many important innovations arisen and been developed largely because of their potential use in waging war. Critics of military-dominated technology claim that it warps the economy, distorts social values, degrades the environment, and threatens life on earth. From the perspective of the late twentieth century, it appears that the military's association with innovative technology may well be one of the hallmarks of our age and the single most significant determinant of the immediate future of the human race.

CHAPTER VI

Selection (2): Social and Cultural Factors

The process by which a novel artifact is selected for replication and inclusion into the life of a people involves various factors, some more influential than others. The previous chapter concentrated on economic and military necessity as selecting agents but in the case of the waterwheel, nuclear reactor, and supersonic transport other forces, notably social and cultural ones, were clearly at work. Ancient and medieval religious beliefs, a bias toward the acceptance of advanced technology, and utopian energy myths each played an ancillary role in the selection of these innovations. In this chapter the social and cultural factors governing selection will be raised to a position of central concern and examined with the aid of cross-cultural comparisons.

Technology and Chinese Culture

The influence of cultural values and attitudes on technological choices is more readily apparent in examples drawn from remote cultures than from those that share our Western outlook. Technology is so intimately identified with the cultural life of a people that it is difficult for an indigenous observer to gain the objectivity necessary for critical appraisal. Fortunately, Chinese history contains a wealth of material on technology and culture, much of which has been studied by modern Western historians. Therefore, it is of Chinese civilization that we first ask the question, How may culture affect the selection and replication of technological novelties?

The three inventions that Sir Francis Bacon identified as the source of great changes in Renaissance Europe – printing, gunpowder, and magnetic compass – were products of Chinese, not European, civilization. According to the English philosopher, this triumvirate

169

was responsible for revolutionizing literature, warfare, and navigation. If these discoveries were of monumental importance in the making of the modern Western world, why did they not exert a similar influence in China? There is no wholly satisfactory answer to this question; and the search for an explanation will take us into an exploration of the cultural values of the Chinese elite.

Printing

The earliest Chinese printing technique was xylography, invented in the eighth century A.D. This technique made use of solid wooden blocks on whose surface the text of an *entire* page was carved. Each incised block was then inked so that repeated impressions could be taken from it on sheets of paper. It was xylography, and not the Chinese invention of movable type (typography) three centuries later, that transformed book production and learning during the intellectual renaissance of the Sung dynasty (960–1279). Through the widespread use of xylography classical philosophical and literary texts were reprinted, authors were stimulated to write new works, library collections grew in size, and generally speaking literacy increased. Apart from their contents, Sung block books embody a level of art and workmanship that remains unsurpassed in the history of Chinese bookmaking.

Paper was first produced in Europe in the twelfth century, one thousand years after its invention by the Chinese, and typography made its initial appearance in Europe sometime around 1440. Unlike paper, the diffusion path of which has been traced accurately from East to West, the route taken by movable type has not yet been identified with certainty. However, evidence exists to support the claim that knowledge of Chinese typography influenced European experimentation with movable type.

The typographical revolution associated with Johann Gutenberg in the fifteenth century was based on cast-metal type and the familiar screw printing press in which the paper was evenly pressed against the inked surface of the assembled pieces of type. The impact of these innovations on Western culture has been celebrated and discussed since the Renaissance. Recent historians have credited printing with the emergence of modern consciousness, the secularization and commercialization of printed matter, the Protestant revolt against the authority of the Roman Catholic Church, the rise of modern science, and the growth of literacy and education.

Gunpowder

Gunpowder, the second of the Baconian heralds of modernity, was first used for military purposes by the Chinese at the beginning of the tenth century A.D. Early Chinese experiments with gunpowder included using it as an incendiary agent and as a rocket propellant for light missiles. Gunpowder-filled explosive grenades were produced in 1231, and by the end of the thirteenth century the Chinese were firing projectiles from guns with barrels made first of reinforced bamboo and later of iron. In less than a century firearms spread from their original home in China to Japan, Korea, the Near East, and finally to Europe.

The earliest drawing of a cannon in Europe dates from ca. 1325. Shortly thereafter artillery was used in warfare and during the second half of the fourteenth century European cannon founders strove to produce large guns capable of hurling heavy projectiles over great distances.

Large European cannons, initially cast in bronze and then in iron, were so effective in destroying the walls of castles and cities that they brought about great changes in the theory and construction of fortifications. Thus the conduct of seige warfare was changed radically by the introduction of gunpowder. On the battlefield, at least prior to the seventeenth century, cannon proved less useful. They were slow to fire and too heavy to be moved about easily during a battle. These drawbacks did not preclude the placement of guns aboard ships. An array of cannon could be mounted on the ship's decks, and the vessel maneuvered in the water to make the best use of its firepower.

When early in the sixteenth century Portuguese ships armed with European-made cannons sailed into Asiatic ports, the defenders found that their antiquated weapons were far inferior to those of the belligerent foreigners. Orientals were forced to agree to the demands for trade and land made by the Europeans, and to weigh the benefits and liabilities of manufacturing Western-style guns for the defense of their sovereignty. The Chinese were placed in the particularly humiliating position of seeking technical help from their intimidators, people whom the Chinese believed were their cultural inferiors. Knowledge of Western military technology was available at a price; the East must adopt those values that enabled the Europeans to manufacture superior guns. But could traditional Eastern societies be preserved if they followed the Western example of aggressively promoting technical change and cultivating modern science? This question plagued the Chinese throughout the 450-year period of Western domination of the Orient, and it continues to trouble the Chinese government today.

The Magnetic Compass

Bacon's last mentioned key invention, the magnetic compass, was first applied to navigation by the Chinese in the eleventh century A.D. In its original manifestation the compass was a divination, or future-predicting, instrument made of lodestone, which is naturally magnetic. This protocompass dates to at least the first century A.D. When the lodestone was replaced by a magnetized needle, about the seventh or eighth century, the compass began to take its more familiar form.

Although the magnetic compass probably became a maritime navigational aid somewhat earlier, we know that by 1080 it was used aboard seagoing vessels and that the Chinese had discovered magnetic declination – the failure of the compass needle to point to true north. Chinese traders were quick to use the compass in opening up new commercial ventures. Early in the twelfth century Chinese seagoing junks, navigating with the help of the compass, carried trade goods to the East Indies, India, and the eastern coast of Africa. Chinese maritime commerce continued to flourish and expand under Mongol rule in the thirteenth and fourteenth centuries.

Directivity, the pole-seeking property of a lodestone, had been known to the Chinese for a millennium before it became the basis of a navigational compass. In the West there was no such prior knowledge of directivity before the twelfth century appearance of the magnetic compass. Although Chinese trading ships regularly used the compass, we cannot be certain that the knowledge of the instrument diffused from East to West. It may have been invented independently in the Mediterranean area.

Whatever its origins, the affect of the magnetic compass in the West was similar to that experienced in China. It made possible long sea voyages out of sight of land and facilitated navigation at night and on cloudy days when the heavenly bodies could not be observed. The compass was one of the technical advances, along with nautical charts, improvements in the design of large sailing vessels, and ship-mounted cannon, that eventually gave European ships command of the major waterways of the earth for nearly five centuries.

Chinese Cultural Stagnation

Having summarized the origins and uses of Bacon's crucial inventions in the East and West, we reiterate the question raised earlier: Why were these discoveries not as influential in changing Chinese culture and technology as they were in the West? A little thought will reveal the inappropriateness and ethnocentricity of this query.

First, it is wrong to assume that the selection of a given novel artifact will have the same meaning and influence in one country as in another, especially when their cultures are so radically different as they were in China and Europe. To imagine that the introduction of printing, to take one example, should generate a precisely similar set of events East and West is a naive viewpoint.

Second, the claim that the Chinese had the essential technological knowledge but that they suppressed it or diverted it into trivial uses is not persuasive. Printing, gunpowder, and the magnetic compass were put to practical uses by the Chinese early, with enthusiasm, and in ways that paralleled ones later developed in Europe.

Third, the attempt to evaluate the comparative impact of identical innovations is grounded in the concerns and values of Western civilization. In effect we are asking why the Chinese are not like us, why they did not initiate the twin revolutions in science and technology that produced our modern world. Or, why did the Chinese not use gunpowder and navigation to rule the seas of the world as we did? Questions of this sort reveal more about the attitudes of the people asking them than they do about the Chinese.

After acknowledging the problems that this line of reasoning raises, there remains the feeling that in some more restricted sense the larger concern is legitimate. Even the strongest defenders of the Chinese as technological innovators admit that there is a great disparity between technology in China and the West in the nineteenth and twentieth centuries. In the fifteenth century, China and Europe were technological equals, granting that Eastern inventions more often flowed from east to west than in the opposite direction. Then came the emergence of modern science in the sixteenth and seventeenth centuries and the establishment of industrial societies in the eighteenth and nineteenth centuries. These were strictly Western phenomena; nothing comparable occurred in the Orient. Moreover, the Eastern civilizations found it very difficult to comprehend, let alone adopt, the manifold results of these massive changes. The West soon became the world leader in science and technology, and the East was left behind.

Scholars have offered several different explanations for the scientific and technological gulf separating East from West. An ingenious, if not always convincing, economic argument has been put forth by historian Mark Elvin. According to Elvin, in the eighteenth century the Chinese economy reached a state that made it incapable of generating and sustaining internal technological changes. Traditional technology had been exploited to its highest level to serve huge markets in China. When there was a local shortage of goods versatile Chinese merchants acted to alleviate the situation by using available means, such as cheap transport, instead of seeking and adopting innovative technological solutions to their problems. In addition, the Chinese economy was so much larger than those of any of the European countries that it would have been impossible to increase it by two- or threefold as was done with the much smaller Western economies. Small-sized economies, more responsive to change and capable of great growth, worked to the advantage of European nations, especially England. China's technological stagnation, therefore, was due to a high-level equilibrium trap that was built into its economy.

Sinologist Joseph Needham singles out the structure of Chinese society and government, and not a static economy, as the source of the profound differences between Chinese and Western technology in modern times. In the third century B.C. China's warring states for the first time were united under a centralized monarchy and the form of government that was established persisted in its essentials until the twentieth century – an imperial government that required large numbers of competent and loyal civil administrators to collect taxes and carry its rule to the far corners of a vast country. Thus was born what has been called Asiatic bureaucratism or bureaucratic feudalism. Entrance to the bureaucracy depended on an extensive knowledge of literary and philosophical classics, notably those by Confucius, and competence was determined by a series of state-sponsored examinations.

According to Needham, the existence of the bureaucratic feudal system mitigated against the rise of a Chinese mercantile class powerful enough to affect governmental policies and actions. By contrast European merchants were in a position to shape social and political decisions and institutions to suit their needs and thus fostered scientific and technological progress.

Everything considered, the absence of a powerful merchant class is a negative argument for a Sinophile to make about Chinese society; therefore, Needham complements it with a more positive one. China's long-lived bureaucratic government, he declares, introduced a stability that was unmatched in Western societies, which continued to be convulsed by recurring social, political, and intellectual revolutions. Given their "steady-state" society, the Chinese were by no means technologically stagnant. They made slow and continuous progress on all scientific and technological fronts for a long period of time until they were overwhelmed by the upheavals in the West. If there is any question that needs to be settled, concludes Needham, it is why Western society and culture was so prone to instability. Whichever explanation we accept, after the social and economic components of the Chinese response to Western science and technology have been studied, we are left facing issues that call for a consideration of prevailing cultural attitudes and values. Even Needham, who offers a strong defense of the socioeconomic approach, admits that hitherto unexamined ideological factors may yet prove to be crucial in explaining China's failure to match the scientific and technological achievements of the Western nations in modern times and its reluctance to adopt the results of those achievements.

The steady-state society that Needham praised can be seen from another perspective as a conservative one bound by traditional Confucian principles, convinced of its superiority over the rest of the world, and suspicious of technological innovations, primarily those coming from the West. In the opinion of some modern historians, the Chinese scholar–officials were men of letters with little interest in, or sympathy for, science, commerce, and utility. Furthermore, the focus on ancient Chinese authors as a field of study was not conducive to an acceptance of the ideas of novelty and progress that had come to the forefront in Renaissance Europe. A Jesuit traveler in the late seventeenth century remarked that educated Chinese were more attracted to antiquities than they were to modern things. He observed that the Chinese predilection for the past directly countered the European's love of novelty for its own sake. Needham's research has shown that the Chinese did have a conception of technical progress; however, their definition and application of the concept was distinctly different from that used by western Europeans.

In addition to being conservative, Chinese society was xenophobic. It was exceedingly reluctant to adopt foreign technologies lest they displace an indigenous, superior mode of life. At this point the interpretation of Chinese behavior and attitudes must be handled with care. A hasty, and oversimplified, explanation is that the Chinese, blinded by their sense of cultural superiority, stubbornly refused to find any merit in the novel artifacts and techniques of alien cultures. A more subtle analysis is that the highly educated officials of Confucian China all too clearly recognized the superiority of Western technology, especially in weaponry. They were willing to risk military defeat by rejecting Western guns and cannon rather than accept them and thus jeopardize the humanistic culture in which they were trained and held power, and which had served as the foundation of Chinese government and ethics for two millennia.

Some Chinese thinkers in the nineteenth century believed that an accommodation could be reached between Eastern and Western ways. They suggested that the Chinese deal with Western technology selectively, carefully separating the artifact itself from those values and customs they found to be repugnant. Wiser men pointed out that the artifact and value system were inseparable. If, to take an example, China adopted European cannon and mechanical clocks, it must necessarily acquire the Western technological methods that made them possible as well as the Western ideas of warfare and time they embodied.

By way of conclusion let us return to Bacon's list of epoch-making inventions and now ask why typography, gunpowder, and the magnetic compass were so readily adopted by Westerners even though the three were products of a remote, foreign land? The answer is that Western culture was not monolithic; Europeans were eclectic, open to new ideas, influences, and things. Because novel artifacts posed no threat to their way of life, Europeans incorporated the Baconian triumvirate into their culture and soon forgot about the innovations's alien origins. By the seventeenth century Francis Bacon could write about these inventions as if they were the results of European, and not Oriental, ingenuity.

Fads and Fashions

At certain points in the development of technology the selection of an innovation is motivated not by widely shared cultural values but by short-lived fads that sweep through a region for a decade or so and are gone. Some of the same selection impulses that we mentioned earlier – enthusiasm for a technological solution, or belief in progress through technology – are operating here too, but where the impulses are associated with a naiveté, fancy, or whimsy, we may conclude that the choice was made on the basis of a passing fad or fashion.

Because cultural values and fads and fashions are located at the opposite ends of the continuum of selecting agents, the former have all too often been dismissed or misinterpreted. Nevertheless, fads and fashion deserve special notice because they too are indicative of values and ideologies that contribute to the development of technology. Furthermore, we must not confuse the strange artifacts that result from a fad moving quickly across the technological scene with the unique, often bizarre, creations of eccentric inventors or mechanics. On the contrary, most fads are produced by established technologists and industries, amply financed by private or governmental sources, and exhibited to the general public. We will focus on just two of these: the atmospheric railway and nuclear-powered vehicles.

Atmospheric Railway

The first steam-powered passenger and freight railroad began operation on 15 September 1830, initiating the railway age in Britain. For the first decade or so construction proceeded at a slow and regular pace as workable roadways and rolling stock were devised, money was raised for laying track, and new routes were planned. Then came the railway mania of the mid-1840s during which 2,800 miles of rails were authorized, more than had been laid in the previous fifteen years. Speculators promoted rail stocks of dubious value, and gullible investors were convinced that the railroad boom would make them rich. Engineer Isambard K. Brunel, himself a builder of railroads, commented on the state of affairs in 1845: "Everybody around seems mad – stark staring wildly mad – the only course for a sane man is to get out of the way and keep quiet."[1]

Although the technology of the atmospheric railway predates the railway-mania years, it was during the time of wild enthusiasm for new rail schemes that the first atmospheric rail companies were founded. Between 1844 and 1847 atmospheric lines were planned or built in England, Ireland, Scotland, Wales, France, Belgium, Austria-Hungary, Italy, and the West Indies.

Atmospheric railway technology is so unlike conventional rail technology that it calls for a detailed explanation (Figure VI.1). The main difference between the two systems is that the atmospheric railway did not use locomotives to pull its trains. Instead, between the tracks of the atmospheric line was laid a cast iron cylindrical tube, fifteen inches or more in diameter, that extended throughout the entire route. A piston, designed to fit snugly in the tube, was fastened to the undercarriage of the leading car of the train. This arrangement required that a continuous, longitudinal slot be cut along the top of the tube so that the bracket connecting piston to railroad car could move freely. To seal the tube, a leather valve was added to keep the slot closed tightly, except when a train was passing through. Therefore, whereas power was supplied in the steam railway by a cylinder and piston situated in the locomotive, the atmospheric

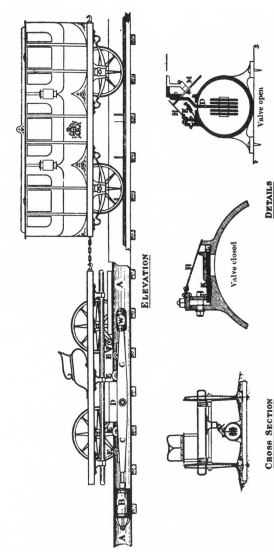

ELEVATION

CROSS SECTION

THE ATMOSPHERIC SYSTEM

A.A. Continuous Pipe fixed between the rails.
B. Piston.
C.C. Iron Plates connected to the piston.
D. Plate connecting Apparatus to Carriage.

E. Metal Rollers to open the Continuous Valve.
F. Roller attached to Carriage for closing the Valve.
H. Weather Valve.[1]

DETAILS

Valve closed

Valve open

K. Continuous Airtight Valve hinged at 1.
L. Composition for sealing Valve.
M. Roller attached to Carriage for opening Weather Valve.[1]
w. Counterweight to Piston.

Figure VI.1. The atmospheric railway. The main technical features of the atmospheric railway system are depicted here. Not shown are the pumping stations that evacuated the air from the continuous cylinder or pipe. Source: Peter Hay, *Brunel: his achievements in the transport revolution* (Reading, 1973), p. 86.

railway located its cylinder at track level, fixed a piston to the first car, and did away with the need for a traction engine.

Another distinctive feature of the atmospheric railway was its use of the pressure of the atmosphere, and not the expansive power of steam, to force the piston through the cylindrical tube. Steam-driven air pumps stationed at two- to three-mile intervals along the rail line were to evacuate air from the cylinder just prior to the arrival of a train, thus causing the piston, along with the train to which it was attached, to move in the direction of the lower pressure.

Because the disadvantages of this innovative railway are readily apparent, let us take notice of a few of its strengths. First, the atmospheric railway offered clean, quiet, and rapid transit to passengers who had experienced the noise and dirt of the early steam railroads. Second, it placed the steam engine, along with its fuel supply, firmly upon the ground. Steam traction wasted energy because the heavy locomotive plus its coal and water had to be hauled over the track continuously. Third, it called for intermittent operation of the steam-driven air pumps. They were needed for about five minutes before the coming of a train; at other times the cylinder need not be emptied of air, which would result in a saving of fuel.

Of the one hundred or so atmospheric railways proposed or built in Great Britain and on the Continent only four were actually completed: one in Ireland, two in England, and one in France. In all a total of 30 miles of atmospheric rail track was laid; the shortest line measured 1.75 miles, the longest 20 miles. The Irish atmospheric railway, which began operating in 1844, was the first to open. The Paris line lasted the longest, opening in 1847 and closing in 1860.

A number of inventors had proposed atmospheric rail schemes in the decades before the 1840s, and the basic technology was established in 1844 after one of the best known engineers in Victorian England, Isambard K. Brunel, became involved. Despite his vow to remain sane in the midst of the railway mania, and despite the adverse reports on atmospheric rail transportation produced by another of the great engineers of the day, Robert Stephenson, Brunel assumed responsibility for building the South Devon atmospheric railway, and the "Atmospheric Caper" began.

The South Devon line traversed some hills, and Brunel had calculated that an atmospheric train, without the extra load of a locomotive, would work best on this terrain. Furthermore, the line would be cheaper to build, he reasoned, because the rails and the roadbed would not have to be strengthened to bear the additional weight of locomotives. After three years of work, the South Devon company opened its atmospheric service to the public on 13 September 1847. One year later the company abandoned its operations. The demise of the South Devon atmospheric line was the single most costly engineering failure of its time and tarnished Brunel's reputation.

The problems encountered on the South Devon and other atmospheric railways were numerous, and in some cases so blatantly obvious that it is hard to understand why first-rate engineering minds failed to foresee them. The absence of a locomotive meant that the driver of an atmospheric train had less control over its movement than did the operator of a pumping station located as much as three miles away. Once the cylinder was evacuated, the piston shot forward and the driver was forced to rely upon a none-too-effective braking system to vary the speed of the throttleless train. The starting and stopping of atmospheric trains, and shunting them from one track to another, also posed great difficulties. The grade crossings at which the traffic of a public road traveled over an atmospheric line with its protruding cylinder presented still other problems.

In theory, fuel was conserved by limiting the operation of the evacuation pumps to a three to five minute period before the arrival of a train. Railroad telegraphy was not yet perfected so the pumping stations were forced to operate on a prearranged schedule. The pumps were in action for much longer than five minutes whenever a train was late and consequently fuel was wasted. The most aggravating and persistent mechanical problems arose from the maintenance of the longitudinal valve on the cylinder. Because speeding trains, moving at fifty or sixty miles per hour, lifted and closed the valve so rapidly, it wore out or broke down frequently. In cold weather the valve froze and refused to close; in hot weather the greasy compound applied to seal the valve melted. When it rained, water leaked through the open valve and the cylinder had to be cleared of it before traffic could move along the line.

To all of this must be added three major flaws in the overall conception of the atmospheric system, flaws that should have been recognized at the outset. First, the atmospheric line relied on a chain of pumping stations. If one of them failed, the entire line was forced to close down; it was impossible to bypass the inoperative pump. Second, the line was restricted to one-way travel. The pumping system could not accommodate travel in two directions simultaneously, and it would have been very expensive to build a parallel second line for traffic moving in the opposite direction. Third, there was no easy way to upgrade the line if traffic on it increased. On a steam railroad one could add a bigger locomotive to pull more cars.

On an atmospheric railroad power was increased only by installing a cylinder that had a larger diameter, adding more powerful steam engines, and adding air pumps of greater capacity at the pumping stations.

How could so many engineers, businessmen, and investors overlook these drawbacks and press on with their ambitious plans for new atmospheric lines? The answer, of course, is that we are dealing with a fad and that those caught up in it were incapable of seeing its many disadvantages. Brunel and the other promoters of the atmospheric railway believed that its serious technical problems would ultimately be resolved. They never were.

Nuclear Propulsion Vehicles

Transportation is a rich field in which to pursue technological fads. Just as interest in the first railroads fostered the atmospheric railway so did the first airplanes spur interest in aircraft designed for personal transportation. During the peak years of America's infatuation with aviation (1900–50), there was much speculation about a family flying machine similar in price, safety, and reliability to the family car. For the suburbanite this would mean an airplane in every garage; for the city dweller, a helicopter on every apartment roof top.

The dream appeared close to realization in 1926 when Henry Ford began to manufacture an airplane that some called the "Ford flying flivver." In the 1930s the Federal Bureau of Air Commerce made its contribution to the fad by financing the design of prototype aircraft that it hoped would be mass-produced like automobiles. One of these, the Plymacoupe, was powered by a Plymouth automobile engine. World War II put an end to such experimentation but not to the dream that the air car would be a common feature of postwar America. Manufacturers promised that the skies would soon be filled with the Skycar, Airphibian, Convaircar, or Aerocar, all capable of traveling on highway or skyway.

None of this came to pass. Personal aircraft remains an expensive, hazardous, and inconvenient means of travel compared with the automobile. Postwar Americans did take to the skies but not in privately owned planes. They flew as passengers in large aircraft piloted and maintained by the professional crews of big corporations.

However futile the search for the poor man's airplane, it was in no way as wasteful as the fad for nuclear propelled vehicles that flourished in the decades immediately after World War II in the United States. Before it had run its course the Federal government spent billions of dollars on nuclear rockets and aircraft and over one hundred million for a nuclear merchant ship.

It had long been a commonplace of the mythology of atomic energy that a pea-sized pellet of uranium contained sufficient energy to lift a freight train to the moon. If that sounded remote and impractical, there was always the prediction that atomic-powered vehicles would provide unlimited travel for a minute expenditure of fuel. Technological developments during World War II appeared to make these forecasts come true. If jet engine technology could be merged with atomic weapons technology, through an effort similar to the Manhattan Project, land, air, and water transportation would be revolutionized. At least this is what the nuclear propulsion faddists believed.

During the 1950s and 1960s enthusiasts who were convinced that the nuclear alternative to conventional technology was always superior obtained over two billion dollars in federal funds for planning and constructing nuclear-powered rockets. In effect these rockets were to be flying reactors in which thrust was produced by air heated in the reactor and then expelled through a rocket nozzle. Nuclear rockets would generate more thrust than a chemical rocket of equal size and could travel over greater distances. The launching of the Russian space satellite Sputnik in 1958 led the nuclear rocketeers to claim that the United States could regain its lost prestige by being the first nation to enter the Nuclear Space Age. Only nuclear rockets, they argued, would permit the Americans to overcome Russian chemical rocket superiority. Under code names Project Pluto, Rover, and Poodle, attempts were made to develop nuclear-propelled rockets. Finally, in 1972 the Atomic Energy Commission decided to terminate funding for these projects because there was no way to neutralize the radioactive exhaust gases.

One need not be a trained engineer to assess the serious problems created by an open-ended reactor spewing radioactive materials onto the ground during launch or into the atmosphere during flight. Of course, this evaluation assumes that the rocket stays aloft once launched. Were it to crash to the earth during launch, or at any time later, the ruptured reactor would release catastrophically high levels of radioactivity.

A far different, but equally hazardous, approach to space propulsion was offered by atomic weapons designer Theodore Taylor and physicist Freeman J. Dyson. They devised a plan for a vehicle capable of moving through space at a hundred thousand miles per hour. The speed was to be achieved and maintained by the timed explosion of a series of nuclear bombs at a short distance behind the

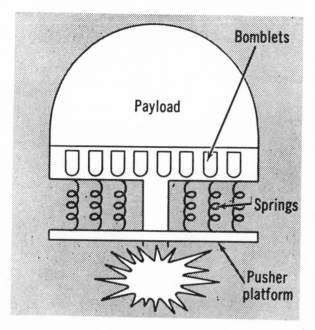

Figure VI.2. Conceptual design for the Project Orion space vehicle. Small nuclear "bomblets," with an explosive force equal to ten metric tons of TNT, explode at regular intervals of one to ten seconds behind the pusher platform. The thrust produced by the explosions is transmitted to the craft by means of water-cooled springs. Source: Joint Committee on Atomic Energy, *Nuclear energy for space propulsion and auxiliary power* (Washington, D.C., 1961), p. 277.

spacecraft, the energies of which would be directed against a huge pusher plate attached to the rear of the vehicle. Thus, the exploding bombs would hurl the spacecraft to the most remote planets in our solar system.

This mad scheme, designated Project Orion (Figure VI.2), cost the government a mere ten million dollars over a seven-year period (1958–65). The Orion design team paid scant attention to the biological and moral implications of introducing nuclear weapons into outer space and polluting the earth's atmosphere with poisonous fallout. Twenty years later in his memoirs (1979) Dyson admitted that the Orion ship was to have been a "filthy creature" that left a "radioactive mess"[2] behind it as it traveled through the cosmos.

Another of the atomic gadgets that fascinated the nuclear propulsion faddists was the nuclear airplane. Shortly after the end of

World War II, work was begun on a reactor propulsion unit that would enable bombers to reach any target on earth and return to the home base without refueling. When this project was finally halted in 1961, more than one billion dollars had been spent and the nuclear airplane was no closer to a final realization than it was in 1948. Once again the problems were insuperable and obvious from the start.

A nuclear jet power plant can be designed in two ways. The first is a direct cycle engine, like the one used in nuclear rockets, in which the air is heated by direct contact with the reactor's fuel elements. An alternative approach is the indirect cycle in which an intermediary material (liquid sodium) transfers the heat from the fuel elements to the air. The latter type reduces the pollution substantially but both reactors call for radically new metallic alloys that can withstand the combined forces of heat and radiation and can resist corrosion.

Even if the reactor materials problem had been solved, there still remained the question of proper shielding. Nuclear rockets had no pilots and hence there was no need to shield the reactor. In a nuclear airplane, it is necessary to protect the crew from the intense radiation of its engine. Large quantities of water or lead would provide effective protection but would add inordinate weight to the plane. What was to be done? One serious response by the aircraft's sponsors was that older crew members be assigned to fly nuclear airplanes because they were less likely to pass on the genetic damage done by radiation to their children. In any case, neither the shielding nor the reactor materials problem was ever solved by the researchers, and consequently the nuclear airplane never left the ground.

The last mode of nuclear transportation subsidized by the U.S. government, the merchant ship, was not a complete technological fiasco. This pioneering effort in maritime nuclear propulsion was conceived as part of President Eisenhower's Atoms for Peace effort. The project's advisors agreed that there was no better way to underscore the program's theme than to adapt the nuclear technology of the submarine for a cargo vessel. With this in mind the President on 30 July 1956 authorized the construction of the *N.S. Savannah*. This ship was to serve as a symbol of the peaceful uses of atomic energy and as a prototype for a fleet of nuclear-powered merchant vessels. After a series of building delays and cost overruns, the *Savannah* was finally ready to depart on her maiden voyage on 31 January 1963.

For the next decade the *Savannah* sailed around the world delivering cargo and passengers. The ship's reactor functioned well, but there were economic difficulties. Originally budgeted at $34.9 million, the *Savannah*'s total costs exceeded $100 million. An additional $3 million was needed annually to cover its losses as a cargo carrier. On the credit side of the account can be recorded the $11 million the ship earned during its eight years at sea. As was the case with the Shippingport generating plant, the nuclear alternative proved to be a very costly one. So costly, in fact, that the promised fleet of nuclear merchant ships was never built.

Fighting ships that can stay at sea for extended periods of time without refueling have an advantage over conventional naval craft. The same does not apply to merchant shipping. Cargo vessels are supposed to move from port to port. As cargo is being loaded and unloaded the vessel can be refueled. It makes no sense to build a cargo ship capable of sailing around the world without ever touching port.

The *Savannah* reached its final berth, Galveston, Texas, in September 1971. There it was decommissioned and permanently mothballed. Built as a symbol of the peaceful atom, the *Savannah* now serves as a reminder of the nuclear propulsion fad that blossomed so exotically and expensively in mid-century America.

Concluding that other facets of present-day technology are somehow immune from the influence of fads would be wrong. At least since the Renaissance, if not earlier, fads and fashions have served as one of the means by which selection is made from among competing novel technological possibilities. Of course it is much easier to identify the fads of the past than to recognize those we have so recently foolishly embraced.

By the mid-1980s the home computer boom appeared to be nothing more than a short-lived and, for some computer manufacturers, expensive fad. Consumers who were expected to use these machines to maintain their financial records, educate their children, and plan for the family's future ended up playing electronic games on them, an activity that soon lost its novelty, pleasure, and excitement. As a result a device that was initially heralded as the forerunner of a new technological era was a spectacular failure that threatened to bankrupt the firms that had invested billions of dollars in its development.

Discard and Extinction

The long-standing bias among historians is to examine the origins of innovation and not their ultimate extinction or discard. Consequently, we know much more about the sources of technological novelty, and the ways novel artifacts are selected, than we do about the process by which a culture divests itself of artifacts that hitherto have served it well.

The complexities of culture's divestment process are nicely illustrated in British anthropologist W. H. R. Rivers's study of the disappearance of utilitarian objects on a number of islands in Oceania. The extinct artifacts – the canoe, pottery, and the bow and arrow – were certainly not occupying marginal niches in the cultures of these South Sea islands, nor were they displaced by "superior" Western equivalents such as motor boats, factory-produced cooking ware, and rifles.

The disappearance of these Oceanic artifacts cannot be explained by a sudden natural catastrophe that destroyed needed raw materials or killed key craftworkers. In the case of the canoe, wood was abundant on the islands, but there is evidence that skilled canoe makers simply died out without leaving any one to carry on their craft. Because this craft was closely associated with magical and religious rites, Rivers suggests that it may well be the perceived loss of spiritual power, or *mana* as it was called, and not the diminution of technical competence that put an end to canoe building.

The disappearance of pottery presents a more difficult case. Not all of the islands have sufficient quantities of raw materials, but potter's clay is available in quantity in the area and pots have been found in ancient grave sites; still contemporary inhabitants do not use pottery. Social factors apparently offer the best explanation. If pottery making was restricted to a few tribes and the wares were traded abroad, then the eradication of these tribes, in wars or through epidemics, would halt pottery production and distribution. As with the canoe, the loss of skills cannot be ruled out as a possible explanation for the extinction of pottery over a wide area.

The bow and arrow never entirely disappeared from Oceania; however, it was put to debased uses, such as target practice or the shooting of rats and birds, instead of serving as a prime military weapon. In warfare the club replaced the bow and arrow, which indicates that the latter's demise may be related to new battle tactics, a different view of the goals of war, or changing attitudes toward death in battle.

From the modern Westerner's perspective, the bow and arrow, pottery, and the canoe are utilitarian objects absolutely necessary for the life and well-being of preindustrial peoples. Yet there were South Pacific islanders who did not share this evaluation. They allowed these three to become extinct not because they had a better alternative at hand, but apparently because the artifacts were in conflict with more powerful social and cultural values.

Extinction cannot be studied apart from the emergence of innovation and its subsequent selection and replication. Invention, replication, and discard, according to material-culture theorist George Kubler, are of equal importance in reaching a better understanding of the made world and how it changes. Invention breaks stale routine, replication makes the invention widely available, and discard assures that there will be room for newly invented things in the future. This interlocking cycle, whose existence is better documented in industrial societies, operates in preindustrial societies as well.

The lack of extensive artifact diversity and the absence of the notion of technical progress in preindustrial societies lead one to assume that the retention of artifacts there is high and that consequently their lifespan, or duration, is long. In a culture where a great deal of effort is expended in making relatively few items, there is an incentive to retain and repair broken things, make do with what one has, and generally preserve the status quo. The pursuit of the untried and the novel is less attractive in these settings in which a great deal of time and energy must be devoted to producing such new things. Because it is far better to accept existing artifacts than it is to innovate, the relative lack of novelty in traditional societies may be due to an understandable reluctance to part with the old instead of an inability to create the new.

If long durations characterize the artifacts of traditional societies, a trend toward shorter duration is what one might expect to find in modern cultures that promote novelty, endorse the doctrine of progress, and cultivate and apply science. To these three powerful factors should be added the techniques of mass production perfected in the first half of the twentieth century. Mass production encourages the discard of individual artifacts (creating a throw-away culture), and the replacement of entire classes of things. This is perpetuated by the ability of those engaged in mass production to replicate an innovation quickly and flood the market with its copies, an act that simultaneously satisfies a craving for novelty, creates a feeling of satiety, and prepares the way for the next innovation. In many cases this process operates at the level of fashion, as with annual automobile model changes, but there are other instances where its influence is by no means superficial. For example, the semiconductor industry's ability to create new and more powerful microchips with regularity is greatly facilitated by its use of the techniques of mass production.

When one class of artifacts replaces another one, the displaced artifacts do not vanish from the scene. For a time there exist over-lapping generations of different artifacts capable of filling, to a degree, the same function. So it was in the 1920s and 1930s when lighter-than-air craft (dirigibles) were being superseded by the air-plane, and so it is with the generations of computers that have succeeded one another so rapidly in the late twentieth century. Not every company immediately buys the latest computer, and if it does the older models are not destroyed but are passed on to other users.

Technological evolution has nothing comparable to the mass ex-tinctions that are of interest to evolutionary biologists. History does not record any widespread, cataclysmic extinctions of entire classes of artifacts, although something similar might occur on a local level in remote communities or on isolated islands. Only in the apoca-lyptic visions of science fiction does one encounter the wholesale destruction of an entire technological civilization and its forced retreat to an earlier, usually paleolithic or medieval, stage of development.

A final feature of artifactual discard that deserves notice is what Kubler calls intermittent duration. An artifact that has been dis-carded is revived and reinstated at a later date. Ignoring the delib-erate revivals of antique technologies by museums for educational and nostalgic reasons, there are cases of discarded artifacts finding a new life under different social, economic, and cultural conditions. Steam locomotives, supplanted in the West by electric and diesel engines, flourish on mainland China where seven thousand of them are in use daily and three hundred new ones are manufactured each year. The revival of the wood stove and solar-heating methods in the 1970s' energy crisis is another example of discarded artifacts given a new life at a later date.

A more striking instance of intermittent duration is the gun in Japan. European firearms were introduced to the Japanese by the Portuguese in 1543. Guns were quickly selected for use in warfare and were produced in large quantities by highly skilled Japanese craftsmen. By the end of the sixteenth century there were more guns, in absolute numbers, in Japan than anywhere else on the globe. At what appears to have been the height of their popularity, however, the Japanese returned to their traditional weapons: the sword, the spear, and the bow and arrow.

There are several reasons why the Japanese did, and could, re-nounce the gun. Japan's elite and influential warrior class, the sa-murai, preferred to battle with swords. The Japanese sword possessed symbolic, artistic, and cultural values that transcended its role as a weapon. It was the embodiment of the warrior's ideas of heroism, honor, and status and was linked to aesthetic theories specifying the proper movements of the human body. The gun, on the other hand, was an alien instrument, devoid of these rich associations. Finally, Japan's insular position and its reputation as a nation of fighters made it possible for the country to rely upon the sword at a time when it was surrounded by gun-using neighbors.

The Japanese never formally abolished firearms; in the seventeenth century government officials simply restricted their use and production forcing gunsmiths to return to sword and armor making or to more mundane metalwork. By the eighteenth century those firearms remaining in Japan were antiquated and largely unused. Japanese military technology and strategy had reverted to the sword as its basic weapon.

So matters stood. Then in 1853 the visit of Commodore Matthew C. Perry led to the opening of Japan to the West and its technology. Following the resignation of the last Tokugawa ruler in 1876, resistance to the Western intrusions collapsed. The Japanese revived the manufacture of firearms and cannon, and the nation was soon on its way to becoming a modern military and industrial power.

In giving up the gun after they had embraced it so enthusiastically and mastered its technology so well, the Japanese proved that deep cultural values can overcome practical considerations. The subsequent revival of firearms, under Western prodding, showed that it was possible to bring back a set of artifacts that had been discarded several centuries earlier. Everything considered, this is an extraordinary case in the history of technology. It serves as a lesson in the renunciation of a key military weapon and dramatically illustrates the processes of intermittent duration and extinction, technological diffusion, and artifact selection.

Alternative Paths

When the Japanese gave up the sword and returned to the gun they were, in effect, following an alternative technological path. And a century later when the United States decided against funding the supersonic transport they too were choosing an alternative possibility – the jumbo jets. All too often it is assumed that the development of technology is rigidly unilinear, that at no point could other choices have been made. This viewpoint is reflected in popular responses to any attempt to limit or otherwise criticize current technological practices. On those occasions we are apt to be told it is impossible to stop or alter the predetermined course of technological progress.

The evolutionary perspective on technological change reveals that there are a diversity of paths open for technological exploration and exploitation. Our investigation into the sources of technological novelty has discovered the many alternatives available, especially in industrial societies. And the study of artifact selection has made clear the arbitrary nature of the decisions made. Again and again neither biological nor economic necessity determined what was selected. Instead decisions were made on the basis of these two elements combined with a large measure of ideology, militarism, fad, and current conceptions of the good life.

In the examples that follow, the branched character of technological evolution will be emphasized. Despite widespread belief that the made world could not be otherwise than it is, in the case of the printing press, railroad, and gasoline engine different choices could have been made. These choices would not necessarily have resulted in a better world but they would have created a livable one that was different to a degree. These three inventions all date from the post-Renaissance period, and their modernity might lead one to argue that choice is only possible with moderate- or short-duration artifacts, that there are few, if any, alternatives in the realm of long-duration objects. Thus, we shall address this issue first.

Hand Tools

Hand tools have a very long duration and among them exists a notable instance of an alternative path in the design and use of a fundamental artifact. Because the hand saw is a tool based on stone prototypes, it is reasonable to assume that the familiar Western saw forms are universal and that people everywhere saw wood in the manner in which it is done in Western Europe and North America. Western saws have a pistol-grip handle that is grasped with one hand as the tool is *pushed* away from the body to make the cut. It turns out that this method of sawing is rather recent, dating from the Roman era. Furthermore, far from being the prevailing way of sawing wood, it is favored solely in the West. In the Orient the hand saw has a straight wooden handle that can be grasped with one or both hands. In use the tool is *pulled* toward the body. The teeth of Eastern saws are raked or slanted in the direction of the handle so that the cut is made on the pull stroke.

There is more at issue here than opposing motor-habit patterns. The tools themselves are different. The handle of the Eastern saw makes it possible to use both of the edges of the blade for cutting. Teeth for rip and crosscut can be on a single instrument. Because

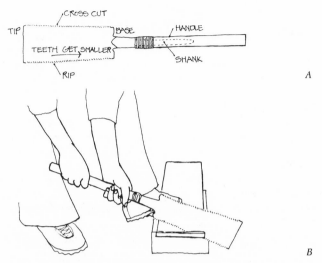

Figure VI.3. A. Ryoba-Nokogiri, standard Japanese carpentry saw used for house building and woodworking. B. Ryoba-Nokogiri in use. A cut across the grain is made by using a two-hand grip on the saw's handle. Source: Kip Mesirow and Ron Herman, *The care and use of Japanese woodworking tools* (Woburn, Mass., 1975), pp. 6, 12.

steel under tension (pulling) is stronger than that under compression (pushing) the blade of the Eastern saw can be quite thin. Blades of Western saws must be relatively thick in order to prevent buckling, bending, and breaking as the tool is thrust into the resisting wood. A thinner blade produces a narrower kerf (or cut) and hence reduces the wood wasted as sawdust. Finally, the two saws are distinct enough so that a worker accustomed to using one of them must be retrained in the operation of the other.

A survey of hand tools around the world reveals that large numbers of pull tools are employed. For example, in China and Japan the wood plane is pulled toward the body rather than pushed away (Figure VI.3). Evidence also suggests that in the recent past even Europeans and Americans made a greater use of pull tools – for example, drawknives, spokeshaves, scrapers, adzes, and scorps to shape chair seats – than they do today.

The conclusion is obvious. If the ancient and simple tools fashioned to carry out the basic operations of woodworking can be designed and used so differently, we can expect to find alternatives for the more complex and adaptable artifacts of industrial cultures.

Block Printing: East and West

From Francis Bacon in the seventeenth century to media theorist Marshall McLuhan in the twentieth century, printing with movable type has been praised as one of the great forces shaping Western thought and life. Economic historian David S. Landes hesitated about putting Gutenberg's invention on the same level as fire and the wheel but was satisfied that it belonged in a second category, along with the mechanical clock. Underlying this high regard for the printing press is the premise that the spread of knowledge in the West or elsewhere is dependent on typography and that no alternative printing technology exists.

The history of events leading up to Gutenberg's achievement is usually presented and interpreted as follows. Before the coming of cast-metal type that could be rapidly assembled to produce any text, there were two methods of reproducing the written word, both of them inadequate. Texts could either be hand copied by scribes, a long, tedious, and error-prone effort, or the entire page could be carved on the face of a single block of wood and an inked impression taken from it. Xylography, or wood-block printing, faithfully reproduced multiple copies, but its greatest drawback was that each page had to be painstakingly carved to order. For this reason xylographic books were expensive and took a long time to produce; they were unsuitable for the diffusion of knowledge to a large audience. Typography made use of mass-produced, interchangeable letters that were assembled for a specific text and disassembled when the printing job was completed. The relationship between xylography and typography was so close that the first movable type might have consisted of letters sawed away from a xylographic block. To conclude this conventional summary of printing history: Xylography was technologically inferior but prepared the way for movable type, which was to have wide cultural repercussions.

Indeed, were it not for a very different set of facts that emerge from the Chinese experience, xylography would appear to be a poor alternative to typography. In the Orient it was xylography that sparked a printing and intellectual revolution and typography that was tried and then dropped because of its shortcomings. Block printing, a Chinese invention of the eighth century A.D., gained prominence two hundred years later with the publication in 953 of the Confucian Classics. This work, consisting of 130 volumes, established the Confucian corpus and restored it to its central place in Chinese literature and thought. The dissemination of Confucianism through the printed page is credited with inspiring a renewed interest in classical learning comparable with the revival of Greek and Roman classics in Renaissance Europe.

interest in classical learning comparable with the revival of Greek and Roman classics in Renaissance Europe.

The Chinese printing renaissance was a xylographic one with block books published on a large scale and in a great variety of titles, covering both secular and official subjects. Dynastic histories, commentaries on the classics, dictionaries, encyclopedias, and local histories were produced in addition to collections of essays and poetry and technical treatises on medicine, botany, and agriculture. In some cases gargantuan publishing projects were undertaken. The entire Buddhist canon, known as the *Tripitaka*, was printed between 972 and 983 in a set of 5,048 volumes covering 130,000 pages. Each page was engraved upon a separate wooden block.

The xylographic books printed in the four centuries between 960 and 1368 are unsurpassed examples of Chinese bookmaking skills. The artistry of the Chinese book had declined somewhat by the fifteenth century; nonetheless, the number of books published continued to increase. Modern authorities on Chinese printing claim that as late as the year 1500 more printed pages existed in China than in the rest of the world. Others, arguing that this is too conservative an estimate, suggest that Chinese printing outstripped the world until 1700 or 1800.

Typographical experiments began in eleventh-century China. Initially the characters were engraved on pieces of soft clay, which was then hardened by baking. Wooden type was tried in China during the thirteenth century, and the Koreans were casting metal type for printing in 1403. Despite these attempts typography was not widely adopted in the Orient. When the Europeans arrived in the sixteenth century, they found xylographic reproduction to be the prevalent means of printing. In the nineteenth century typography was reintroduced into the land of its origins by Westerners.

How can we account for the utter rejection of typography in the Orient? The twofold answer to this question combines the aesthetic with the practical. As an art form, typographic books never achieved the excellence of their xylographic competitors. This difference meant a great deal to a people who cultivated calligraphy as an art and who were sensitive to the nuances of book design. As a practical matter, those who espoused typography had to handle at least five thousand different Chinese characters in printing, a source of various problems that contributed to the unpopularity of typography in the East.

There still remains the troublesome problem of the time and effort needed to engrave xylographic blocks. A modern student of printing commissioned a contemporary wood engraver to carve a block for a single, small, page measuring five by seven inches. The engraver took thirty to thirty-five hours to complete the task. Yet, the sixteenth-century Italian missionary to China, Matteo Ricci, who was well acquainted with Western printing techniques, noted that a Chinese artisan could carve an entire page in about the same time that it took for a European typesetter to set a folio-sized page with metal type. In support of Ricci's observations, there are the millions of xylographic books that were distributed throughout the vast Chinese empire for centuries.

The conventional account of printing, recounted at the opening of this section, suggested that one of the incentives to the invention of movable type was the difficulty encountered in preparing the xylographic blocks. From that we can draw two inferences: that xylography existed in the West prior to Gutenberg's day, and that copies of pretypographic block books are extant in European libraries and museums. However, according to the best available information, *no* Western block books predate Gutenberg. The earliest xylographic books date from 1460, at least two decades after movable type is said to have been invented. And block books continued to be printed in Europe for about a century.

In the West xylography remained a vastly underdeveloped technology. Instead of acting as a disseminator of ideas, it was used to popularize biblical stories, relate simple moral tales, reprint common prayers, and summarize the rudiments of grammar. Instead of serving as the highest artistic model of the art of bookmaking, block books in the West were cheap, crudely printed and illustrated, and usually very short. The longest known block book in Europe contains ninety-two leaves printed on both sides.

The wide discrepancy between Eastern and Western xylography cannot be explained in terms of the written languages of the two regions. If Chinese artisans could engrave some five thousand intricate ideograms for very high-quality books why could not European craftsmen learn to engrave the twenty-six simple letters of their alphabet with proficiency and artistry? It is easier to understand why the Chinese shunned typography than it is to determine why xylography was of so little consequence in Europe.

The study of xylography in the West is an esoteric subject limited to a small number of specialists. The general understanding of it has been hindered by the myths that have surrounded typography and by Western ignorance of Chinese xylographic accomplishments. From a technological viewpoint, at least one thing is clear: The block book could have met the needs of Renaissance Europe just as it did those of Sung Dynasty China. This is not to say that the

West need never have adopted typography but to argue that xylography was a viable alternative that certainly could have served the West for several centuries.

Railroads versus Canals

Between 1840 and 1960 journalists, economists, and professional historians agreed that the railroad was the single most influential determinant of economic growth in nineteenth-century America. The railroad revolution was credited with the westward advance of agriculture, the rise and form of the modern corporation, the development and siting of industry, the establishment of patterns of urbanization, and the structure of trade between the major regions of the country. In 1891 Union Pacific railroad president Sidney Dillon could claim that the welfare of the American people depended upon the nation's rail system. "No imagination can picture," he wrote, "the infinite sufferings that would at once result to every man, woman, and child in the entire country"[3] should the railroads be destroyed. Seventy years later economic historian Robert W. Fogel dared to do the unthinkable; he removed the railroad from nineteenth-century America and assessed the consequences. The absence of the railroads, he concluded, would not have greatly affected economic growth between 1840 and 1890 because canal and river boats, with the help of horse-drawn wagons, could have moved goods normally carried by rail and because the railroad was not of vital importance to the market for manufactured items or as a stimulus for technological innovation.

Before recapitulating Fogel's argument, his assumptions and aims should be clearly understood. A cliometrician, or quantitative economic historian, Fogel believes that historians should not only be concerned with past events but also with their alternative possibilities. To this end he created a model of nineteenth-century America's economy without railroads and compared this counterfactual model to the economic reality of the period. His sole criterion for judging the effects of the railroad was its impact upon economic growth. To test the common knowledge that railroads were an indispensable ingredient in the development of the economy, Fogel determined whether alternative means of transportation might have replaced the rail system.

One of the main claims of the proponents of the railroad revolution was that the interregional distribution of agricultural products could only have been accomplished by long-distance rail transportation. They argued that the hauling of foodstuffs from farm lands to urban centers not only aided industrialization but encouraged the settlement of the Midwest. In short, the railroad made large areas of the country economically accessible and contributed to overall economic growth.

Fogel noted that the westward movement of the population, in its initial phases at least, was not dependent upon the railroad. By 1840 about 40 percent of the American people lived west of New York, Pennsylvania, and the coastal Southern states, yet no railroads extended from the East into those newly populated regions. Settlers who made the westward trek relied heavily on natural waterways and canals. Furthermore, they began large-scale farming even though there were no railroads to carry their goods to Eastern markets. Michigan, Ohio, Kentucky, Tennessee, Indiana, Illinois, and Missouri together produced 50 percent of the national corn crop in 1840, a time when those states had a mere 228 miles of unconnected railroad track in operation. The corn was shipped by water, as was cotton from the inland states to the South. In 1860, 90 percent of all cotton delivered to New Orleans was sent by barge or boat.

Navigable waterways and wagons supplied adequate transport for people and goods in the early part of the century, but could they have handled the increased traffic at a later date? After all, the population continued to grow and land that was more remote was cultivated. Surely something on the order of the railroad would have been needed by then.

Fogel's response to these doubts was a careful geographical study of all commercial farm land with the aim of determining its access to useful waterways. He found that a large percentage of the land was located on an average of forty miles, as the crow flies, from a river or canal and that if five thousand miles of additional canals had been built in Illinois, Iowa, and Kansas, then 93 percent of the agricultural land served by railroads would have been within reach of a canal or river. Had improvements been made on existing public roads, the accessible acreage would have risen to a higher percentage. The programs of canal extension and road improvement called for by Fogel were well within the technological and economic capabilities of nineteenth-century America.

Even if Fogel's entire analysis is correct, the alternative technological path he suggests still might have been an expensive one to follow. In calculating the relative costs of rail and water transportation, with allowance for the freezing of northern canals in winter, the slower movement of goods by canal boat, and the need for frequent transshipment of canal freight, Fogel determined that the railroad offered a small advantage over the canal in the interregional shipment of agricultural products. However, the difference – less than 1 percent (0.6 percent) of the gross national product in 1890 – is scarcely sufficient to support a theory of the revolutionary importance of American railroads in the latter half of the nineteenth century.

Agricultural land in America was spread out over every state in the union and yet the greater part of it was, with the help of wagons, accessible to water transport. Because deposits of iron ore and coal were not so widely distributed, the railroad might well be needed in the regions where they were mined. But again Fogel's survey of the major mining areas convinced him that the railroad was not indispensable in the hauling of coal and ore because the mines were located close to waterways.

This summary of Fogel's work does not do justice to the ingenuity and rigor of his argument or to the vast amounts of data he assembled to support his case. Also absent are strong negative responses from his critics. Although many points remain open to debate, Fogel has seriously challenged those who argue that the invention of the railroad was inevitable and a major contribution to progress in the nineteenth century. He appropriately emphasizes canals, which provided the first inexpensive and efficient means of transportation. Canal barges offered a 90-percent decrease over the prevailing wagon-hauling rates and even charged slightly lower rates than did the railroads. Although the canal would not have remained competitive with the railroad in all areas forever, the canal clearly provided an alternative mode of transporting agricultural goods and raw materials in the nineteenth century.

Steam, Electric, and Gasoline Vehicles

While speculating about the state of transportation technology, Fogel also hypothesized that in a nonrail America vehicles powered by internal combustion engines might have been developed earlier than they were. He thought that some portion of the millions of dollars invested in railroads could have been diverted to the creation of an alternative mode of transportation. The time that elapsed between the understanding of the principles of the internal combustion engine in the early nineteenth century and its embodiment in a working model in the 1860s might have been shortened were there no railways.

Throughout the nineteenth century attempts were made to produce a supplement, if not exactly an alternative, to the railroad, including a steam-powered road vehicle. Inventors in England,

France, and the United States separately devised self-propelled steam cars and buses that were large, heavy, difficult to maneuver, and mechanically unreliable. These steam carriages, built to travel over the poorly maintained public roads of the day, could not compete with the railroad whose wheels rolled on smooth, hard metal rails. By the end of the century the steam vehicle experimenters had produced lightweight, powerful engines that, along with electric motors and gasoline engines, were used to power the first generation of automobiles.

At the turn of the century, it was by no means obvious that the modern automobile engine – the Otto four-stroke cycle internal combustion engine – would win out over its competitors (Figure VI.4). In 1900, 4,192 cars were manufactured in the United States. Of them, 1,681 were steam, 1,575 electric, and only 936 gasoline. Shortly thereafter, however, the internal combustion engine began its climb to prominence. At the New York automobile show in 1901, 58 steam, 23 electric, and 58 gasoline models were put on display. By 1903 the number of steam and electric models exhibited had fallen to 34 and 51 respectively, whereas the gasoline cars numbered 168. At the 1905 exhibit the rout was complete; the 219 gasoline models displayed outnumbered by a ratio of 7 to 1 the combined total of steam and electric cars. Unfortunately, it is much easier to document the triumph of the internal combustion engine than it is to explain its success. In 1905 each of the power plants had advantages and disadvantages; none had a clear-cut technological superiority.

The electric car appeared to have all of the good points of the horse and buggy with none of its drawbacks. It was noiseless, odorless, and very easy to start and drive. No other motor vehicle could match its comfort and cleanliness or its simplicity of construction and ease of maintenance. Its essential elements were an electric motor, batteries, a control rheostat to regulate speed, and simple gearing. There was no transmission and, hence, no gears to shift.

The first commercial electric vehicles were produced in 1894. Within five years of their appearance on city streets, Henry Ford's boss at the Edison Illuminating Company in Detroit was urging Ford to stop wasting his time tinkering with gasoline engines. Electricity, he argued, would provide energy for the car of the future. Inventors Elmer Sperry and Thomas Edison, who agreed with this prediction, worked on their own versions of the electric automobile. If, as many believed, the twentieth century was destined to be the electrical age, then there was no place in it for the noisy, exhaust-spewing internal combustion engine.

A

B

C

Figure VI.4. Three American automobiles dating from the late 1890s. These automobiles are remarkably similar in appearance yet their powerplants differed radically: A. Baker electric; B. Stanley Mobile steamer powered by a two-cylinder steam engine; C. Autocar powered by a two-cylinder internal combustion engine. Source: Albert L. Lewis and Walter A. Musciano, *Automobiles of the world* (New York, 1977), pp. 82, 83, 78. Courtesy The Conde Nast Publications, Inc.

The electric car was not without serious faults. It was slow, unable to climb steep hills, and expensive to own and operate. Above all else, it had a limited cruising range. Its heavy lead and acid storage batteries had to be recharged every thirty miles or so. The electric was not a vehicle in which to tour the countryside or drive to a distant city. The installation of battery-charging stations in Boston, Philadelphia, and New York was designed to facilitate urban travel but did not solve the problem of long-distance driving. That solution called for lighter and more powerful batteries, a goal that continues to elude electric car promoters to this day.

Because the electric vehicle's restricted operating range met the requirements of urban delivery service, electric trucks were built for the movement of goods within the city. Department stores, bakeries, and laundries purchased them as did the American Railway Express Company. Initially, electric trucks proved economical but by the mid-1920s they were displaced by delivery vehicles powered by gasoline engines, which were less expensive to purchase.

Steam automobiles enjoyed intense popularity at the beginning of the century. To understand this phenomenon, we must set them apart from the cumbersome steam carriages of the previous century. The power plant in an early twentieth-century Stanley or White steamer was a trim twenty to thirty horsepower unit, about the size of a gasoline engine, made of precision-machined steel parts, and fueled with a petroleum product. The overall appearance of a steamer was similar to that of a gasoline motor car of the period.

The steam car was not as quiet as the electric but its purchase price and upkeep were considerably less and a powerful engine enabled it to handle all road conditions without strain. The first self-propelled vehicle to reach the summit of Mount Washington, New Hampshire, was a Stanley Steamer (1899), as was the first car to travel over two miles a minute (1906). The steamers of the 1900s could outstrip the electric car but did face competition from the best of the gasoline engine automobiles.

A look at the engines of the steamer and the gasoline-powered cars points out significant differences. The steam engine's ability to deliver maximum power as it revolved at a slow and steady rate was an important factor in its success. Whereas the reciprocating internal combustion engine in gasoline-powered cars ran at 900 revolutions per minute (rpm) when idling and 2700 rpm at maximum efficiency the engine on the steamer at a speed of sixty miles per hour revolved at a slow 900 rpm. On internal combustion engines an elaborate set of gearing (transmission) was necessary to transmit and transform this rotative power so that it could move the wheels at an acceptable speed, but the steamer had no need for a transmission, clutch, and gearshift. The timing, cooling, valving, and carburation of the internal combustion engine all demanded special attention in its design and added to the number of moving parts; the steam engine had far fewer moving parts than did the gasoline engine, which meant less engine wear and easier maintenance. Finally, the use of proper fuel was critical for the gasoline engine, but low-grade petroleum distillates could be burned to heat the water of the steamer.

The steamer did have a number of important drawbacks. The limited cruising range that plagued the electric also proved troublesome for the steamer. Because steam was exhausted to the atmosphere, and not condensed for reuse, the steamer needed a water refill every thirty miles. Another problem was the time it took to generate steam for the car's first run of the day, although the standard half-hour wait was eventually reduced to a few minutes with the introduction of pilot lights and flash boilers. Steam-engine builders, well aware of these shortcomings, were seeking ways to overcome them.

Steam-automobile manufacturing lasted through the 1920s and ever since then there have been periodic rumors of its revival. The most recent call for the rebirth of steam power was heard in the 1970s, a time of concern over automobile emission pollution and a petroleum shortage. How reasonable are these calls for a steam car revival? Could the steam engine have powered America's great automobile revolution of the twentieth century? Why did the steam engine fail to meet the challenge of the internal combustion engine?

These questions become more, not less, difficult to answer when the internal combustion engine of the 1900s is evaluated. Early gasoline motor cars were rather clumsy and complex machines. To start their engines, a hand crank and muscle power were necessary; their successful operation depended on a series of recently fashioned mechanical systems for ignition, cooling, lubrication, and transmission of power, and they were noisy and emitted unpleasant exhaust gases. On the positive side the gasoline motor car did have an extended cruising range of seventy miles, offered reliable if not trouble-free transportation, and could climb most hills and travel at a good speed on the road. Furthermore, technicians were perfecting the gasoline engine by improving its ratio of horsepower to weight. Everything considered, steam and gasoline automobiles were not that radically different in the transportation they provided.

The selection of the gasoline engine was not the result of a rational appraisal of the merits of the competing power plants. There were no automotive experts at the turn of the century, only inventors and entrepreneurs following their hunches and enthusiasms and trying to convince potential car owners to buy their product. Given this situation, once the gasoline engine gained ascendancy, steamers and electrics were either forgotten or viewed as missteps along the road to automotive progress. Thereafter, money, talent, and ideas were invested in improving the internal combustion engine. Few were willing to champion the steam engine and fewer still to finance its improvement.

Steam and gasoline automobiles were the real contenders; the electrics had the reputation of being a car for the well-to-do and had a seemingly intractable battery problem. The steam engine suffered because of its identification with the technology of the previous century. Steam power called to mind huge locomotives or stationary engines belching black smoke, burning tons of coal, and periodically bursting their boilers. Steam scarcely seemed suitable as a motive power for a new century. In terms of modernity, electricity would have been the ideal choice, but, if it could not be had, then the internal combustion engine seemed preferable to an updated version of the steam engine.

The one hundred or more makers of steamers did little to overcome the steam engine's negative images and thereby popularize their vehicle. The Stanley brothers, the most successful of the steam car builders, lacked the ambition and managerial skills needed to produce cars in quantity and distribute them across the nation. Nor were they quick to incorporate existing technical improvements that would have made their vehicles more attractive to customers.

In 1914 Henry Ford visited the Stanley factory, which was then turning out 650 cars annually; Ford was manufacturing that many of his Model Ts in a single day. While skilled craftsmen slowly built and hand-finished a few Stanley steamers, unskilled workmen on Ford's innovative assembly lines were mass-producing thousands of gasoline automobiles. The Stanley Company weathered the restrictions placed by the government on the American auto industry during World War I, but it emerged in a weakened condition. Shortly after the war, the company closed its doors; it had failed to meet the competition of the inexpensive Detroit motor car.

Some historians have cited geographical factors as the cause of the rise of the gasoline automobile. Steam and electric cars were primarily built and sold in the eastern United States. The gasoline car, on the other hand, was particularly well-suited to the rural areas of the Midwest. The midwestern predilection for gasoline cars coincided with the industrial and natural resources of the region. Its ample supply of hard woods had earlier made the Midwest a center of carriage and wagon production, and its farm power supply needs had attracted makers of stationary gasoline engines. Therefore, the Midwest could easily supply the main elements – body and engine – of the new gasoline automobile when it became a popular form of transportation.

Is there nothing more to be said about the contest between the steam and gasoline engine except that some shrewd midwestern businessmen chose the latter and used their entrepreneurial skills to make it the basis for the nation's personal transportation system? Up to this point we have not mentioned the relative efficiencies of the two engines because it was not isolated as a separate issue at the time. However, the study of theoretical and actual heat engines indicates that an Otto cycle engine is superior in thermal efficiency to a steam engine. Thus, everything else being equal, a gasoline engine yields more miles per gallon of fuel than a steam engine. Here, at last, is solid evidence that the promoters of the gasoline engine were on the right track. Consciously, or intuitively, they had backed the most efficient engine.

Apart from the fact that the thermal efficiency of the rival power plants was lost in the plethora of technical and cultural factors that favored the gasoline engine, there is yet another problem. The same engineering sources that report the superior thermal efficiency of the Otto cycle engine also disclose that the diesel engine is far more efficient than the gasoline engine. Under actual driving conditions, the average thermal efficiency of the Otto cycle is about 10 percent and a diesel 18 percent. Therefore, if the early Detroit entrepreneurs forsook the inefficient steam engine, why did not later, and better informed, automotive engineers lead the country to more efficient diesel power? The answer, of course, is that the selection of automobile engines, whether in the early or late twentieth century, is made on other than purely technical and economic grounds.

A counterfactual world in which the steam engine powered automobiles and trucks is every bit as reasonable as one in which canal boats moved heavy goods cross-country or xylography served as the basis for a printing revolution. The contest between the steam and gasoline engines was a much closer one than we have been led to believe, so close that under different conditions the steam engine might have won. In America, where petroleum was cheap and plentiful, transportation could have been powered by external combustion rather than internal combustion engines.

It is not an accident that two of the three examples of alternative technologies surveyed here are taken from modern industrial societies. Nathan Rosenberg has noted that such societies are not utterly dependent on a single innovation because, if necessary, they are capable of generating substitutes for it. If Rosenberg is correct, and there is good reason to think he is, then the examples discussed here are not unique. Alternatives can be found not only for railroads and gasoline engines but for almost any major modern invention. The production of novelty is so great that clusters of related innovations, waiting to be selected, exist to fulfill virtually any of our wants, needs, or whims. The history of technology would be written far differently if, instead of concentrating on the "winning" innovations perpetuated by selection and replication, we were to make a diligent search for viable alternatives to those innovations.

Conclusion

Implicit in the discussion of the selection process in Chapters V and VI is the assumption that the selecting agents are active, productive individuals capable of making the choices and changes needed to shape the material world as they see fit. The selectors do not represent all segments of society nor are they necessarily concerned with the public's welfare. However, they have the freedom to decide which of the competing novelties will be replicated and incorporated into cultural life. Some restrictions are placed on this decision-making activity, but opportunities for change are plentiful.

This *voluntaristic* approach to technological change, so named because it assumes that humans have the freedom and will to act effectively, is opposed by a group of philosophers and social critics who favor a deterministic explanation of change. In the words of Langdon Winner, a prominent modern spokesman for the group, "The idea that civilized life consists of a fully conscious, intelligent, self-determining populace making informed choices about ends and means and taking action on that basis is revealed as a pathetic fallacy."[4] The impossibility of self-determination is not the result of the machinations of a powerful ruling elite but is due to the nature of twentieth-century technology. The leading question is not *who* but *what* governs society? Winner's answer is "autonomous technology," technology changing in accordance with its needs and not the needs, desires, or wants of humankind.

According to Winner the voluntaristic outlook was developed before the late nineteenth century at a time when hand tools and a small number of machines were the dominant technological instruments and when it was still possible to alter the course of technology. One set of tools could be exchanged for another, or an entire class of machines replaced by another class better suited to meet society's goals. The selection of the waterwheel, steam engine, and reaper fall into this category.

The freedom to develop technology primarily to serve human needs was lost with the spread of industrialization and the growth of modern megatechnical systems in communications, transportation, power production, and manufacturing. These gigantic, complex, interconnected technological systems overwhelm human values and defy human control. Change is possible in the system only if it does not conflict with primary technical values such as efficiency or large-scale integration. Hence, the way we live, work, and play is structured by the monolithic technological order that rules modern industrial society.

A specific example might help to clarify Winner's claim that the megatechnical systems we have created now dominate us. The electrical system that provides energy for domestic and industrial light, heat, power, and communications serves the needs of tens of millions of people. However, the system that produces and distributes electricity is so large and complex, and we are so dependent upon it, that our first concern must be its maintenance as a functioning entity. A power blackout of only a few hours duration paralyzes entire sections of the country; a prolonged power outage creates social chaos. Therefore, we cannot undertake any radical changes in our electrical system for fear of disrupting its technological integrity. Men and women may sit at the control panels of electric generating plants and on the governing boards of electric power companies, but their freedom of action is restricted by the technological master they serve. The controllers and governors can maintain the status quo, guard against the system's deterioration and destruction, and make changes certain to increase the system's operating efficiency, but they are unable to reorganize it or replace it with a different system.

Does the idea of autonomous technology as developed by Winner and others conflict with the selection process discussed here? There are three reasons for answering negatively. First, the idea of technology out-of-control has been criticized for not accurately reflecting the state of large-scale technology. Modern men and women are not helpless victims of the technological order. Second, even if we accept the most extreme form of technological determinism, there is still room for change, albeit change that is in complete harmony with the technical demands of the system and not with social needs. Third, a less rigorous formulation of autonomous technology acknowledges the existence of very powerful megatechnical constraints but gives the selector some freedom of choice. This modified or attenuated determinism is similar to what was found in the case of the supersonic transport. The aircraft industry's half-century drive for increased airspeed was deflected by social, economic, and political forces; at the last moment, a decision was made that overrode the purely technical demand. In sum, for the purposes of a theory of technological evolution, the selecting agents need not enjoy total freedom of action, with all possible choices equally available. It will suffice to have a narrower range of choices from competing novelties and a restricted field of operation for the selector.

CHAPTER VII

Conclusion: Evolution and Progress

Evolution

It is fitting that a book based on an evolutionary model should conclude, as it began, with reference to the work of Charles Darwin. Although Darwin never considered applying his evolutionary theory to technology, a number of Darwin's contemporaries readily drew analogies between the development of living beings and material artifacts. The earliest, and perhaps most famous, nineteenth-century figure to do so was Karl Marx, who published his *Capital* in 1867, eight years after the appearance of Darwin's *Origin of Species*. Marx's evolutionary analogy includes two stages. In the first stage technology engages humanity in a direct, active relationship with nature. Men and women use their labor to shape physical reality, thus creating the artifactual realm. Once the natural world is transformed by work, nature becomes a virtual extension of the human body. Thus, men and women working with natural objects and forces bring nature within the sphere of human life.

Having minimized the differences between the made and the living worlds, Marx moves on to the second stage of his argument and suggests that the Darwinian approach to the "history of Nature's Technology" be transferred to the "history of the productive organs of man."[1] He argues that evolutionary explanations should be applied to the organs that plants and animals rely upon for survival *and* to the technological means that humans use to sustain life. Given his assumption that important features of the human body can be explained in evolutionary terms, then so too can technology, the body's extension into nature. There are, however, significant differences between Marxian and Darwinian evolution. In Darwin's theory biological evolution was self-generating; in the Marxian scheme the

207

evolution of technology is not self-generating but is a process directed by willful, conscious, active people and molded by historical forces.

Neither Marx nor any of the others who attempted to explain the development of technology along Darwinian lines in the nineteenth and twentieth centuries used the available historical data and scholarship to work out the full implications of the evolutionary analogy. That enterprise, neglected for so long, is central to this book.

The concept of diversity, which stands at the beginning of evolutionary thinking, is basic to an understanding of technological evolution. An appreciation for the rich diversity found in the made world has been dulled by our familiarity with the products of technology and hampered by our unquestioning acceptance that those products are absolutely essential for our survival. Artifacts are uniquely identified with humanity, – indeed they are a distinguishing characteristic of human life; nevertheless, we can survive without them. That is why José Ortega y Gasset declared in 1933 that technology was the production of the superfluous. Fire, the stone axe, or the wheel are no more items of absolute necessity than are the trivial gadgets that gain popularity for a season and quickly disappear. Biological necessity is not the reason that so much thought and energy are expended on the making of novel artifacts. People make new kinds of things because they choose to define and pursue human life in this particular manner. The history of technology is not a record of the artifacts fashioned in order to ensure our survival. Instead, it is a testimony to the fertility of the contriving mind and to the multitudinous ways the peoples of the earth have chosen to live. Seen in this light, artifactual diversity is one of the highest expressions of human existence.

If artifactual diversity is to be explained by a theory of technological evolution, then we must be able to demonstrate that continuity exists between artifacts, that each kind of made thing is not unique but is related to what has been made before. Artifactual diversity has inspired our search for evolutionary explanations, and continuity is the first prerequisite for explanations of that sort. A theory of evolution cannot exist without demonstrated connections between the basic units that constitute its universe of discourse. In technology those units are artifacts.

The prevalence of artifactual continuity has been obscured by the myth of the heroic inventive genius, by nationalistic pride, by the patent system, and by the tendency to equate technological change with social, scientific, and economic revolutions. However, once we actively search for continuity, it becomes apparent that every novel artifact has an antecedent. This claim holds true for the simplest stone implement and for machines as complex as cotton gins and steam engines; it applies to inventions dependent upon scientific research and theory, like the electric motor and the transistor; and it pertains to large-scale technological systems as well as to the fantastic machines in works of science fiction. Whenever we encounter an artifact, no matter what its age or provenance, we can be certain that it was modeled on one or more preexisting artifacts.

In the early nineteenth century French astronomer Pierre-Simon de Laplace postulated the existence of a Divine Intelligence who, knowing the velocities and exact positions of all of the atoms in the universe at a given moment, would be able to determine the history of the physical universe and accurately predict its future. Modifying his proposal somewhat, I could argue that given a similar omniscient intelligence, one that was capable of knowing the antecedents of all existing and vanished artifacts, it could reconstruct the grand and vast network of linked artifacts that constitute the history of material culture. Such reconstruction would reveal innumerable streams of related artifacts converging on that remote point in time when the first object was shaped by protohuman hands. In tracing the artifact streams back in time, problems arise only when we try to determine the sources of novelty and account for the differential rate of the appearance of novelty among past and present inhabitants of the earth.

History and the social or psychological sciences provide inadequate explanations for the appearance of novel artifacts within the made world. Therefore, in this book I have explored a number of the major sources of novelty without formulating a comprehensive theory to explain its emergence. The importance of play and fantasy in the creation of technological innovation has been stressed here because their significance has been overlooked by scholars who believe necessity is the sole spur to invention. Knowledge, either as scientific research or in the form of artifacts and technical understanding transferred from one culture to another, has long been recognized as a source of novelty. Despite the importance of technological knowledge, its precise role in technological diffusion and its relationship with scientific theory has only begun to be studied by historians.

Socioeconomic and cultural factors that stimulate innovation certainly deserve our attention. The supporting scholarship for the economic interpretation of innovation is notable for the amount of data it has amassed and the ingenious reasoning it has advanced. Yet, in the final analysis, the arguments put forth on behalf of the economic interpretation are unconvincing, and we are obliged to seek out those cultural factors that fuel the drive for novelty.

As unsatisfactory as any of these concepts may be, it is important to remember that because novelty is a fact of material culture, the evolutionary theory developed here remains intact despite our inability to account fully for the emergence of novel artifacts. Modern theorists of technological evolution indeed face the same dilemma that confronted the Darwinists in 1859. The latter could point to reproductive variability as a fact of nature but were unable to explain precisely how and why variants arose because they did not possess a knowledge of modern genetics. We who postulate theories of technological evolution likewise have our Darwins but not our Mendels.

The process by which novel artifacts are selected from among competitors for incorporation into the culture of a people is an aspect of technological evolution that interests both historians and social critics. Technological alternatives have been evaluated and selected throughout history by a process that continues to operate, despite rising pressures to restrict the freedom of technological choice. Economic and military necessity, social and cultural attitudes, and the pursuit of technological fads have all influenced the selection of novel artifacts. Understanding the nature of those influences may help us to make better-informed decisions in the future. At times socioeconomic and cultural restrictions have limited the search for technological alternatives, and we should clearly recognize the source of the limitations. All too often we have been told that technical constraints impose restrictions on our freedom of selectivity when the fault lies with powerful sociocultural attitudes or institutions.

It is extraordinarily difficult for us to imagine an alternative technological world, especially one strikingly different from our own and yet not inferior to it. European and American proponents of progress have long taught that existing Western technology is superior to all others and that it can only be improved by technological advances made in the West at some future time. Yet the historical analysis of printing, of canals and railroads, and of competing power plants for automobiles raises serious questions about the validity of this outlook. We need instead to take a close look at the relationship between technological and human progress.

Technological Progress

The concept of technological progress, which has shaped thinking about the nature and influence of technology since the Renaissance, is based on six assumptions. First, technological innovation invariably brings about a marked improvement in the artifact undergoing change; second, advancements in technology directly contribute to the betterment of our material, social, cultural, and spiritual lives, thereby accelerating the growth of civilization; third, the progress made in technology, and hence in civilization, can be unambiguously gauged by reference to speed, efficiency, power, or some other quantitative measure; fourth, the origins, direction, and influence of technological change are under complete human control; fifth, technology has conquered nature and forced it to serve human goals; and sixth, technology and civilization reached their highest forms in the Western industrialized nations. These assumptions are encapsulated in the well-known example of the advent of steam power. In the nineteenth century cultural commentators claimed that the inventors of the steam engine gained control of powerful natural forces, raised the amount of energy available per capita, and enabled their fellow citizens to reach a new stage in civilization based on increased energy consumption.

Opposition to the idea of progress appeared as early as the seventeenth century, but not until the mid-twentieth century did all six assumptions come under strong critical attack on a variety of issues. Modern warfare proved that death and destruction were undisputable fruits of technological advancement and that the increase in available energy made possible by the splitting of an atom's nucleus not only failed to create an advanced form of civilization but threatened existing social and cultural attainments, if not all life on earth. The growing awareness of the deleterious ecological by-products of technological expansion revealed that human subjugation of nature was far from secure and that it was gained at the expense of severely polluting the environment. Finally, the long-held belief in the inherent superiority of Western technology was challenged by those who cogently argued that some non-Western technologies better served human needs without disrupting the natural world.

As proponents of progress found it increasingly difficult to present the control of nature or the betterment of human life as the goal of technological advance, they redoubled their efforts to use physical quantities as indicators of technological progress. Thus they argue, even if we cannot prove that modern men and women have been more successful than their ancestors in taming nature or creating a better life, we can at least agree that proof of progress can be found in the fact that modern land vehicles move more rapidly than ancient models or that modern agricultural methods produce larger crop yields than earlier ones. At first glance these claims appear to be self-evident, but under careful scrutiny such "objective" measures of progress can be shown to be as vulnerable to criticism as the subjective ones.

For example, the increase in speed of land vehicles throughout history has been repeatedly offered as incontrovertible evidence of the progress of transportation technology. At one extreme (ca. 5000 B.C.) is the sledge, moving over the ground on wooden runners at speeds of 1 or 2 miles per hour, and at the other the 1983 Rolls-Royce jet-engine racing car capable of traveling more than 630 miles per hour. Between these extremes are animal-drawn carts, carriages of various kinds, and vehicles powered by steam, electric, and internal combustion engines. These conveyances can be arranged hierarchically according to their maximum velocities, and from the assembled data we can draw an ascending curve by plotting vehicle speed and historical time. Such a curve, which has been used to demonstrate the reality of technological progress, assumes that land transportation is a fixed human need that exists outside of any cultural context. Prehistorian V. Gordon Childe responded to this evidence with a critical question: "Did a reindeer hunter in 30,000 B.C., or an Ancient Egyptian in 3000 [B.C.], or an Ancient Briton in 30 [B.C.], really need or want to travel a couple of hundred miles at 60 m.p.h.?"[2] Childe's points are well taken: Human needs are constantly changing, and the speeds of land travel appropriate to one time and culture are not necessarily appropriate to another. For the twentieth-century historian to place the ox-cart and the motor truck side by side on a modern highway and measure their respective maximum speeds as evidence of technological progress simply will not do. Modes of land transportation, like any other technology, must be evaluated in terms of the cultures in which they were conceived and used. Cross-cultural comparisons, or those made within a given culture over extended periods of time, are very poor sources of data on which to rely for establishing the advancement of technology.

Some critics might argue that our example is too extreme and that land transportation needs change over time and from culture to culture, whereas the need for food remains relatively stable. Therefore, let us select two cultures that grow and consume grain – one using primitive agricultural methods and the other modern ones – and resume the search for signs of progress. Grain yield per acre comes to mind as a meaningful way to assess the relative efficiencies of the two methods of food production. The modern farmer, as might be expected, clearly emerges as the winner in this contest.

If we compare Mexican cut/burn (also known as slash/burn) agriculture, in which axe and hoe are used to clear and work the land, with contemporary American agriculture, in which machines and chemicals are used to clear and work the land, the yield of corn by weight is 2.8 times greater on the American farm. But this does not reflect the energy expended in each of the two methods and introduces a bias; high crop yields of the American farm depend on large quantities of energy expended in the forms of agricultural chemicals, fuel, and machinery. By contrast the Mexican farm crop is raised with not much more energy than is supplied by the men and women actually working the land. When the energy output of the crops is compared with the energy input of the farming practice, the result is startlingly different. On the Mexican farm the energy output/input ratio for corn is 11:1 and on the American farm it is 3:1. The lesson to be drawn here is not that the cut/burn method is superior and should be adopted by American farmers but that grain yield is a deceptive measure and that agricultural practices must be evaluated in terms of specific cultural attitudes and needs, rather than by some supposedly neutral quantitative test.

Given the problems raised by the subjective and objective measures of technological progress, is it still possible to think of technology as advancing toward a fixed goal? V. Gordon Childe believed so. In a series of books written in the midst of economic depression and war (1936–1944), he sought to justify the idea of human progress through technological advances. Admitting that the national and international crises of the day forced one to be pessimistic about the progress of humanity, he argued that the historian who takes a long view will find good reasons to be optimistic.

Childe's outlook was based on the material evidence of prehistoric archaeology and the theories of evolutionary biology. He believed that conventional history, limited in evidence, scope, and time, must be replaced by a historical understanding that joined prehistory to recorded history and recognized the importance of the material culture of preliterate peoples. From his perspective the periods of stagnancy and degeneracy that characterize recent history are minor anomalies in the forward thrust of humankind. The retarding effects of war, poverty, famine, and the like are more than compensated for by humanity's positive accomplishments: the manufacture of stone implements, the creation of agriculture, the building of the first urban centers, the working of metals, and the invention of writing. To Childe archaeological evidence of these achievements constituted tangible proof of human progress.

By expanding the modern historical vision to incorporate prehistory, Childe brought the study of history closer to archaeology and to zoology, paleontology, and geology. His investigation into the technical accomplishments of the earliest human beings raised questions about the human evolutionary path and the relationship between biology and culture. According to Childe the biological record provided evidence of the development of apelike creatures into the first humans and the archaeological record revealed the continuous improvement made by humanity since its initial appearance on earth. Using an explicit evolutionary analogy, he likened changes in human culture to the modifications and mutations that gave rise to new species of animals; he suggested that what the historian called *progress* was known to the zoologist as *evolution*.

Biology gave Childe what he called "the final test of progress,"[3] a numerical test that was scientific and devoid of the value-laden metaphysical outlook that had dominated discussions about human progress for centuries. The evolutionary principle of the survival and propagation of plant and animal species implied that a species' fitness or success could be accurately measured by counting the total number of members of a species that survived from generation to generation. If that number was increasing, Childe judged the species was fit or successful; if not, he listed it as a failure. Given this approach, the human animal was an integral part of the evolutionary process, and its fitness, or ability to progress, could be determined by the same test of population growth that had been applied to all other organisms. When cultural and technological changes led to increases in the human population, Childe said that the changes were progressive and that humankind had advanced.

Childe believed that he had found unassailable proof of a correlation between technological change and population increase in the events associated with England's Industrial Revolution. Between 1750 and 1800 the curve describing England's population growth turned sharply upward, a development that Childe explained by reference to the sweeping technological, economic, and social changes that historians have termed the Industrial Revolution. According to Childe, historical debate about the benefits and horrors of industrialization could finally be resolved by recourse to his objective criterion. He concluded that the Industrial Revolution was a progressive event in history because it facilitated the survival and multiplication of the species it affected most – homo sapiens.

Having settled arguments over the impact of the Industrial Revolution to his satisfaction, Childe turned to his primary concern, prehistory. Because he had demonstrated that the Industrial Revolution had accelerated human progress, he would also be able to demonstrate that a number of great prehistoric revolutions led to an increase in population. The revolutions that gave humankind agriculture, metals, cities, and writing each resulted in the proliferation of the human species. Therefore, the Industrial Revolution was not a unique event; it was merely the latest in a series of technological revolutions that had hastened humankind's progress through the ages.

Childe believed that the population test liberated the idea of progress from "sentimentalists and mystics"[4] and placed it squarely in the camp of the scientist. Yet his measure of progress is not above criticism. It was a brilliant stroke to enlarge the scope of the discussion of human progress to include prehistoric times. Local disturbances that loom large in conventional historical accounts recede as the historian studies the entire record of humanity. Nonetheless, prehistory has its own set of problems. Historical evidence for the prehistorian is limited to the material remains that have survived the accidents of time. Hence, little is known about prehistoric technologies other than lithic and ceramic ones and virtually nothing about the lives and thoughts of prehistoric people. From this skewed and scanty record, entire human cultures must be reconstructed, a project bound to yield a vague and highly conjectural picture of the past. How can one claim that prehistoric societies progressed when so little is known about them? Childe's answer is that we at least know whether their populations increased or decreased and that is sufficient information for the determination of progress.

The connections Childe made between progress in prehistoric and modern times depended on his assertion that England's population grew dramatically between 1750 and 1800 and that the growth was to be explained by technological advances. This assertion is no longer acceptable to modern demographers. Some argue that the increase in population occurred before 1750 and should be listed as but one of the causes of the Industrial Revolution. Others accept a later upturn in population growth, which they attribute to the vagaries of weather or disease rather than to industrialization. Because historians disagree about the compilation and interpretation of late eighteenth- and early nineteenth-century demographic data, it is not surprising that they also disagree about the effects of a series of events that supposedly took place at a remote time when no written records were kept. Even if we had reliable information on population changes in the prehistoric period, it would still be necessary to prove that any population increase noted was the result of technological advancement.

Other flaws plague Childe's population test and deserve to be mentioned, if only in passing. Two are critically important. First, if proliferation is a sign of progress then humans, with a world population of 5×10^9, trail behind several more prolific organisms. For example, some species of insects are said to number more than 10^{16} individuals, and *Euphausia superba*, the small krill eaten by whales, has a population greater than 10^{20}. Second, in the 1930s Childe equated population growth with success; yet less than a half-century later human proliferation has raised fears that overpopulation will strain our living space, environment, food supply, and natural resources. Population increase is not counted as an unqualified good in a world in which the inhabitants of overpopulated regions experience the lowest standard of living.

Our rejection of Childe's population test necessitates a return to the highly polarized debate over the impact of technology on society. At one pole are enthusiasts who praise the wonders of technology and science and urge that the pace of technological change be hastened so that civilization can be advanced more rapidly. At the opposite pole are environmentalists and social critics who stress the ill effects of technological growth and argue for restraints. These conflicting viewpoints, which are vigorously defended by their respective supporters, defy easy resolution. Nevertheless, we can formulate a modified concept of technological progress that is compatible with the opposing positions and with an evolutionary theory of technological change.

A modification of the idea of technological progress requires that two fundamental alterations be made in the traditional view. First, progress in technology must be determined within very restricted technological, temporal, and cultural boundaries and according to a narrowly specified goal. Second, the advancement of technology must be disengaged from social, economic, or cultural progress. As an illustration of my approach, let us return to the case of radio wave transmission discussed in Chapter III. Between 1887 and the early 1900s several researchers worked to increase the distance over which electromagnetic radiation could be transmitted. German physicist Heinrich Hertz, who was primarily concerned with establishing the existence of James Clerk Maxwell's hypothetical radiation, was satisfied to transmit across the limited fifteen-meter space of his laboratory. In 1894 Oliver Lodge demonstrated transmissions of fifty-four meters at the Oxford meeting of the British Association for the Advancement of Science. Then Guglielmo Marconi arrived in England, determined to make radio telegraphy a reality by sending signals over much greater distances. In 1894–95 he began with transmissions of a few hundred meters; by 1899 he was able to communicate by radio across the English Channel; and in 1901 he achieved the first transatlantic radio communication.

This series of steady increases in the transmission distance of electromagnetic radiation constitutes technological progress. The events took place within a limited time span, less than twenty years, and within a relatively homogeneous cultural setting, England and Germany. The goal was simply the transmission of a radio signal over ever greater distances. From Hertz to Marconi, the basic transmitting technology remained constant – the generation of radio signals by means of intermittent sparks created by an induction coil or a bank of capacitors. Because the intermittent spark transmission of Morse code was replaced by the continuous wave transmission of the human voice in the 1920s, my determination of progress has been confined to the period well before 1920. Finally, no reference has been made to radio transmission and the advancement of civilization. The governing assumption has been that if a group of people find it desirable to transmit coded radio signals, then increases in transmission distance can be seen as signs of progress for that particular technological undertaking.

Similar examples of technological progress can easily be found in the history of technology: increasing the power of the atmospheric steam engine prior to its redesign by Watt; reducing the exposure time for the photographic plate used in the first daguerreotype cameras; maintaining the Wright brothers' biplane aloft for longer periods of time than the fifty-seven seconds of the initial flight at Kitty Hawk; or keeping Thomas Edison's carbon filament incandescent light bulb glowing brightly for an extended time. These examples, all drawn from the early stages of inventions, can be supplemented by evidence provided by mature, modern technologies such as reducing automobile engine knock in the 1920s; improving the image transmitted via television between 1930 and 1939; or miniaturizing a specific transistor in the late 1950s.

The reformulation of technological progress should be acceptable to the opponents in the technology-versus-society debate and still satisfy the condition that evidence for progress not be gathered across technological and cultural boundaries and over long periods of time. But most important, all of these examples accord with the evolutionary analogy that is central to this book. Organic evolutionists since Darwin have been reluctant to accept the idea of life evolving toward some predetermined goal. Mention of direction, purpose, or progress in connection with organic evolution is avoided because it is thought to introduce metaphysical speculation into scientific discourse. In similar fashion I have resisted the tendency to make the advancement of humanity or biological necessity the end toward which all technological change is directed. Instead, I explain artifactual diversity as the material manifestation of the various ways men and women throughout history have chosen to define and pursue existence. Although choices are consciously made to fulfill immediate goals, such as flight in a heavier-than-air craft or increased fuel efficiency for an automobile engine, the sum total of those choices does not constitute human progress.

A workable theory of technological evolution requires there be no technological progress in the traditional sense of the term but accepts the possibility of limited progress toward a carefully selected goal within a restricted framework. Neither the historical record nor our understanding of the current role of technology in society justifies a return to the idea that a causal connection exists between advances in technology and the overall betterment of the human race. Therefore, the popular but illusory concept of technological progress should be discarded. In its place we should cultivate an appreciation for the diversity of the made world, for the fertility of the technological imagination, and for the grandeur and antiquity of the network of related artifacts.

Bibliography

The items listed here are keyed to specific chapter sections so that the reader can quickly determine which books and articles are relevant to the topic being discussed. There are, however, a number of general histories of technology that deserve special mention. These books will satisfy the reader's need for a traditional account of the development of technology. Multivolumed histories of technology include Charles Singer, et al., eds., *A history of technology*, 7 vols. (Oxford, 1954–78); Maurice Daumas, ed. *A history of technology and invention*, 3 vols., trans. E. B. Hennessy (New York, 1969–79); Melvin Kranzberg and Carroll W. Pursell, Jr., eds., *Technology in Western civilization*, 2 vols. (New York, 1967). For single-volume treatments of the subject see T. K. Derry and Trevor I. Williams, *A short history of technology from the earliest times to A.D. 1900* (New York, 1961); Trevor I. Williams, *A short history of twentieth-century technology c. 1900–c. 1950* (New York, 1982); D. S. L. Cardwell, *Turning points in Western technology* (New York, 1972); Arnold Pacey, *The maze of ingenuity* (New York, 1975); Abbott Payson Usher, *A history of mechanical invention* (Cambridge, Mass., 1954).

I. DIVERSITY, NECESSITY, AND EVOLUTION

Diversity

Ernst Mayr, *Evolution and the diversity of life* (Cambridge, 1976); Thomas J. Schlereth, *Material culture studies in America* (Nashville, 1982); Karl Marx, *Capital*, ed. F. Engels, trans. S. Moore and E. Aveling, vol. 1 (New York, 1972); e. e. cummings, "pity this busy monster, manunkind," in *Poems, 1923–1954* (New York, 1954);

David S. Miall, ed., *Metaphor, problems and perspectives* (Sussex, 1982); Earl R. MacCormac, *A cognitive theory of metaphor* (Cambridge, 1985).

Necessity

Aesop, *Aesop's fables: a new edition with proverbs and applications* (London, 1908); James J. Flink, *The car culture* (Cambridge, Mass., 1975); Robert F. Karolevitz, *This was trucking* (Seattle, 1966).

The Wheel

David S. Landes, *Revolution in time: clocks and the making of the modern world* (Cambridge, Mass., 1983); Wilfred Owen, Ezra Bowen, and the Editors of *Life*, *Wheels* (New York, 1967); Stephen Jay Gould, "Kingdom without wheels," in *Hen's teeth and horse's toes* (New York, 1983); Stuart Piggott, *The earliest wheeled transport* (Ithaca, N. Y., 1983); M. A. Littauer and J. H. Crouwel, *Wheeled vehicles and ridden animals in the ancient Near East* (Leiden, 1979); Lazlo Tarr, *The history of the carriage*, trans. E. Hoch (New York, 1969); Gordon F. Ekholm, "Wheeled toys in Mexico," *American Antiquity* 11 (1946), 222–8; Richard W. Bulliet, *The camel and the wheel* (Cambridge, Mass., 1975).

Fundamental Needs

Bronislaw Malinowski, *A scientific theory of culture* (New York, 1960); Philip Steadman, *The evolution of designs* (Cambridge, 1979); Benjamin B. Beck, *Animal tool behavior: the use and manufacture of tools by animals* (New York, 1980); José Ortega y Gasset, "Man the technician," in *History as a system* (New York, 1961); Gaston Bachelard, *The psychoanalysis of fire*, trans. Alan C. M. Ross (Boston, 1964).

Organic–Mechanical Analogies

L. J. Rather, "On the source and development of metaphorical language in the history of Western medicine," in *A celebration of medical history*, ed. Lloyd G. Stevenson (Baltimore, 1982); Charles Webster, "William Harvey's conception of the heart as a pump," *Bulletin of the History of Medicine* 39 (1965), 508–17; René Descartes, *Treatise of man*, trans. Thomas S. Hall (Cambridge, Mass., 1972); Samuel Butler, *Erewhon or over the range*, ed. Hans-Peter Breuer and Daniel F. Howard (Newark, 1980); Samuel Butler, "Darwin on the origin of species," "Darwin among the machines," "Lucubratio Ebria," and "The mechanical creation," in *The works of Samuel Butler: Canterbury Settlement* (New York, 1968); Hans-Peter Breuer, "Samuel Butler's 'The Book of the Machines' and the argument from design," *The Journal of Modern Philology* 72 (1975), 365–83; Patricia S. Warrick, *The cybernetic imagination in science fiction* (Cambridge, 1980); Geoff L. Simons, *The biology of computer life* (Boston, 1985); A. Lane-Fox Pitt-Rivers, *The evolution of culture and other essays* (Oxford, 1906); M. W. Thompson, *General Pitt-Rivers: evolution and archaeology in the nineteenth century* (Bradford-on-Avon, 1977).

Cumulative Change

Karl Marx, *Capital*, vol. 1, trans. Samuel Moore and Edward Aveling (New York, 1967), p. 372; William Fielding Ogburn, *Social change* (New York, 1922); William Fielding Ogburn, *On culture and social change: selected papers*, ed. Otis D. Duncan (Chicago, 1964); S. C. Gilfillan, *Inventing the ship* (Chicago, 1935); S. C. Gilfillan, *The sociology of invention* (1935; repr. Cambridge, Mass., 1970); Abbott Payson Usher, *A history of mechanical inventions* (Cambridge, Mass., 1954); Vernon W. Ruttan, "Usher and Schumpeter on invention, innovation, and technological change," *Quarterly Journal of Economics* 73 (1959), 596–606.

II. CONTINUITY AND DISCONTINUITY

Introduction

H. G. Barnett, *Innovation: the basis of cultural change* (New York, 1953), pp. 227–30, 242–3; Brooke Hindle, *Emulation and invention* (New York, 1981); Devendra Sahal, *Patterns of technological innovation* (Reading, Mass., 1981).

Science, Technology, and Revolution

Thomas S. Kuhn, *The structure of scientific revolutions* (Chicago, 1970); Gary Gutting, ed., *Paradigms and revolutions: appraisals and applications of Thomas Kuhn's philosophy of science* (Notre Dame, Ind., 1980); Imre Lakatos and Alan Musgraves, eds., *Criticism and the growth of knowledge* (Cambridge, 1970); Karl R. Popper, *Objective knowledge* (London, 1972), pp. 256–84; Everett Mendelsohn, "The continuous and the discrete in the history of science," in *Constancy and change in human development*, ed. Orville G. Brim, Jr. and Jerome Kagan

(Cambridge, Mass., 1980), pp. 75–112; John Krige, *Science, revolution, and discontinuity* (Sussex, 1980); Robert J. Richards, "Natural selection and other models in the historiography of science," in *Scientific inquiry and the social sciences*, ed. Marilynn B. Brewer and Barry E. Collins (San Francisco, 1981), pp. 37–76; I. Bernard Cohen, *Revolution in science* (Cambridge, Mass., 1985); I. Bernard Cohen, *The Newtonian Revolution* (Cambridge, Mass., 1980), pp. 3–51; I. Bernard Cohen, "The eighteenth-century origins of the concept of scientific revolution," *Journal of the History of Ideas* 37 (1976), 257–88; Don Ihde, "The historical–ontological priority of technology over science," in *Existential Technics* (Albany, N.Y., 1983); Martin Heidegger, *The question concerning technology and other essays*, trans. William Lovitt (New York, 1977); Theodore A. Wertime, "Man's first encounter with metallurgy," *Science* 146 (1964), 1257–67; D. S. L. Cardwell, "Science, technology and industry," in *The ferment of knowledge*, ed. G. S. Rousseau and Roy Porter (Cambridge, 1980), pp. 480–1; Edward W. Constant II, *The origins of the turbojet revolution* (Baltimore, 1980), pp. 1–32; Derek J. De Solla Price, "Is technology historically independent of science? A study in statistical historiography," *Technology and Culture* 6 (1965), 553–68; Edwin T. Layton, "Mirror-image twins: the communities of science and technology in 19th century America," *Technology and Culture* 12 (1971), 562–80; Brooke Hindle, *Technology in early America* (Chapel Hill, N.C., 1966).

Case Studies in Continuity

Stone Tools

André Leroi-Gourhan, "Primitive societies," in *A history of technology and invention*, vol. 1, ed. Maurice Daumas, trans. by E. B. Hennessy (New York, 1969), pp. 18–58; Jacques Bordaz, *Tools of the Old and New Stone Age* (New York, 1970); H. H. Coghlan, "Metal implements and weapons," in *A history of technology*, ed. Charles J. Singer, et al. (Oxford, 1954), pp. 600–22; Robert F. G. Spier, *From the hand of man: primitive and preindustrial technologies* (Boston, 1970), pp. 21–39.

The Cotton Gin

Anthony Feldman and Peter Ford, *Scientists and inventors* (London, 1979), pp. 92–93; John W. Oliver, *History of American technology* (New York, 1956), pp. 132–3; Mitchell Wilson, *American science and invention* (New York, 1954), pp. 78–81; Jeannette Mirsky and

Allan Nevins, *The world of Eli Whitney* (New York, 1952), pp. 66–7; Constance M. Green, *Eli Whitney and the birth of American technology* (Boston, 1956), pp. 45–9; Charles A. Bennett, *Roller cotton ginning developments* (Dallas, 1959); Charles S. Aiken, "The evolution of cotton ginning in the southeastern United States," *Geographical Review* 63 (1973), 196–224; André Haudricourt and Maurice Daumas, "The first stages in the utilization of natural power," in *A history of technology and invention*, vol. 1, ed. Maurice Daumas, trans. E. B. Hennessy (New York, 1969), pp. 103–4; Maureen F. Mazzaoui, *The Italian cotton industry in the later Middle Ages, 1100–1600* (Cambridge, 1981), p. 74; Joseph Needham, *Science and civilization in China*, vol. 4, pt. 2 (London, 1965), pp. 122–4; Kang Chao, *The development of cotton textile production in China* (Cambridge, Mass., 1977), pp. 76–80; Charles C. Gillispie, ed., *A Diderot pictorial encyclopedia of trades and industry*, vol. 1 (New York, 1959), pl. 34; Daniel H. Thomas, "Pre-Whitney cotton gins in French Louisiana," *The Journal of Southern History* 31 (1965), 135–48; Grace L. Rogers, "The Scholfield wool-carding machines" in *Contributions from the Museum of History and Technology, Papers 1 to 11* (Washington, D.C., 1959), pp. 2–14; Charles A. Bennett, *Saw and toothed ginning developments* (Dallas, 1960); Douglas C. North, *The economic growth of the United States: 1790–1860* (New York, 1966), p. 8; Robert Brooke Zevin, "The growth of cotton textile production after 1815," in *The reinterpretation of American economic history*, ed. Robert W. Fogel and Stanley L. Engerman (New York, 1971), pp. 122–47; George Kubler, *The shape of time: remarks on the history of things* (New Haven, Conn., 1962).

Steam and Internal Combustion Engines

E. P. Thompson, *The making of the English working class* (New York, 1966), p. 190; L. T. C. Rolt, *Thomas Newcomen* (London, 1963); H. W. Dickinson, *A short history of the steam engine* (London, 1963); Joseph Needham, *Clerks and craftsmen in China and the West* (Cambridge, 1970), pp. 136–202; Maurice Daumas and Paul Gille, "The steam engine," in *A history of technology & invention*, vol. 3, ed. Maurice Daumas (New York, 1979), p. 45; C. Lyle Cummins, Jr., *Internal fire* (Lake Oswego, N.Y., 1976), pp. 1–182; Aubrey F. Burstall, *A history of mechanical engineering* (Cambridge, Mass., 1965), pp. 332–9.

The Electric Motor

L. Pearce Williams, *Michael Faraday* (New York, 1965), pp. 151–8; W. James King, *The development of electrical technology in the 19th*

Century: the electrochemical cell and the electromagnet, U. S. National Museum, Bulletin 228 (Washington, D.C., 1962), pp. 260–71; Robert C. Post, *Physics, patents, and politics* (New York, 1976), pp. 74–83; Malcolm MacLaren, *The rise of the electrical industry during the nineteenth century* (Princeton, 1943), pp. 87–8; Howard I. Sharlin, *The making of the electrical age* (New York, 1963), pp. 173–5.

The Transistor

Friedrich Kurylo and Charles Susskind, *Ferdinand Braun* (Cambridge, Mass., 1981), pp. 27–9; D. G. Tucker, "Electrical communications," in *A history of technology*, vol. 7, pt. 2, ed. Trevor I. Williams (Oxford, 1978), pp. 1230–48; Ernest Braun and Stuart Macdonald, *Revolution in miniature* (Cambridge, 1978); Charles Weiner, "How the transistor emerged," *IEEE Spectrum* 10, no. 1 (1973), 24–33; G. L. Pearson and W. H. Brattain, "History of semiconductor research," *Proceedings of the IRE* 43, no. 12 (1955), 1794–806; "The transistor," *Bell Laboratories Record* 26, no. 8 (1948): 321–4; Stuart Macdonald and Ernest Braun, "The transistor and attitude to change," *American Journal of Physics* 45 (1977), 1061–5.

Edison's Lighting System

Thomas P. Hughes, *Networks of power: electrification in Western society: 1880–1930* (Baltimore, 1983), pp. 27–9; Harold C. Passer, *The electrical manufacturers: 1875–1900* (Cambridge, Mass., 1953); Harold C. Passer, "The electric light and the gas light: innovation and continuity in economic history," *Explorations in Entrepreneurial History* 1 (1949), 1–9; Matthew Josephson, *Edison* (New York, 1959), pp. 175–267; Robert E. Conot, *A streak of luck* (New York, 1979), pp. 117–201; Christopher S. Derganc, "Thomas Edison and his electric lighting system," *IEEE Spectrum* 16, no. 2 (1979), 50–9; Brian Bowers, *A history of electric light & power* (London, 1982), pp. 141–4.

Barbed Wire

Frank Hole and Robert F. Heizer, *An introduction to prehistoric archeology* (New York, 1973), pp. 220–1; D. S. L. Cardwell, "The academic study of the history of technology," in *History of science*, ed. A. C. Crombie and M. A. Hoskins, 7 (1968): 114; J. Bucknell Smith, *A treatise upon wire; its manufacture and uses* (New York, 1891), pp. 1–98, 312–35; Walter P. Webb, *The Great Plains* (Boston, 1931), pp. 270–318; John J. Winberry, "The Osage orange: a botanical artifact," *Pioneer America, The Journal of Historic American Material Culture* 11 (1979), 131–41; Jesse S. James, *Early United States barbed wire patents* (Maywood, Ill., 1966); Robert T. Clifton, *Barbs, prongs, points, prickers, & stickers: a complete and illustrated catalogue of antique barbed wire* (Norman, Okla., 1970); Earl W. Hayter, "Barbed wire fencing – a prairie invention," *Agricultural History* 13 (1939), 189–217; C. Boone McClure, "History of the manufacture of barbed wire," *Panhandle-Plains Historical Review* 23 (1958), 1–114; Henry D. McCallum and Frances T. McCallum, *The wire that fenced the West* (Norman, Okla., 1965).

A Book-Writing Machine

Lemuel Gulliver [Jonathan Swift], *Travels into several remote nations of the world*, vol. 1 (London, 1726), pp. 71–5; Ann Cline Kelly, "After Eden: Gulliver's (linguistic) travels," *English Literary History* 45 (1978), 33–54; Irvin Ehrenpreis, "Four of Swift's sources," *Modern Language Notes* 70 (1955), 98–100; Hugh Plat, *The jewel house of art and nature* (London, 1653), pp. 42–3; John Locke, *Some thoughts concerning education*, 4th ed. (London, 1699), pp. 272–3; *A new English dictionary on historical principles*, (Oxford, 1888–1933), s.v. "Die."

The Origins of the Discontinuous Argument

Paolo Rossi, *Philosophy, technology and the arts in the early modern era*, trans. Salvator Attanasio (New York, 1970); Edgar Zilsel, "The genesis of the concept of scientific progress," *Journal of the History of Ideas* 6 (1945), 325–49; Louis C. Hunter, "The heroic theory of invention," in *Technology and social change in America*, ed. Edwin T. Layton, Jr. (New York, 1973), pp. 25–46; David A. Hounshell, "The inventor as hero in America," unpublished ms. (1980); Hugo A. Meier, "Technology and democracy, 1800–1860," *Mississippi Valley Historical Review* 43 (1957), 618–40; Asa Briggs, *The age of improvement* (London, 1959); L. Sprague De Camp, *The heroic age of American invention* (New York, 1961); J. B. Bury, *The idea of progress* (New York, 1932), pp. 324–33; John F. Kasson, *Civilizing the machine* (New York, 1976), pp. 152–3; Samuel Smiles, *Lives of the engineers*, 5 vols. (London, 1874-99); Eugene S. Ferguson, "Technical museums and international exhibitions," *Technology and Culture* 6 (1965), 30–46; Matthew Josephson, *Edison* (New York, 1959),

p. 222; Charles Susskind, *Popov and the beginnings of radiotelegraphy* (San Francisco, 1962); V. M. Skobelev, "Incandescent lamp," *Great Soviet encyclopedia*, vol. 14, pp. 92–3; Carleton Mabee, *The American Leonardo: a life of Samuel F. B. Morse* (New York, 1943), pp. 309–11; Gordon Hendricks, "The Edison motion picture myth," in *Origins of the American film* (New York, 1972), pp. vii–xvii, 1–216; Raymond Williams, *Keywords* (New York, 1976), pp. 137–8; Friedrich Engels, *The condition of the working class in England*, trans. W. O. Henderson and W. H. Chaloner (Oxford, 1971).

Conclusion

Michael Partridge, *Farm tools through the ages* (Reading, 1973); Merrill Denison, *Harvest triumphant: the story of Massey-Harris* (New York, 1949), p. 121; L. J. Jones, "The early history of mechanical harvesting," in *History of technology*, vol. 4, ed. A. Rupert Hall and Norman Smith (London, 1979), pp. 101–48; Siegfried Giedion, *Mechanization takes command* (New York, 1969), pp. 146–62.

III. NOVELTY (1): PSYCHOLOGICAL AND INTELLECTUAL FACTORS

Introduction

H. G. Barnett, *Innovation: the basis of cultural change* (New York, 1953); John Jewkes, David Sawers, and Richard Stillerman, *The sources of invention*, 2nd ed. (London, 1969); Arnold Pacey, *The culture of technology* (Cambridge, Mass., 1983); Abbott Payson Usher, *A history of mechanical inventions* (Cambridge, Mass., 1954); *Encyclopedia Britannica*, 14th ed., s.v. "Material culture" by H. S. Harrison; S. C. Gilfillan, *The sociology of invention* (Cambridge, Mass., 1970); William F. Ogburn, *On culture and social change*, ed. O. D. Duncan (Chicago, 1964); Hugh G. J. Aitken, *Syntony and spark – the origins of radio* (New York, 1976); Raymond Firth, *Primitive Polynesian economy* (New York, 1950).

Fantasy, Play, and Technology

Johan Huizinga, *Homo ludens: a study of the play element in culture* (New York, 1970); Jacques Ehrmann, ed., *Game, play, literature* (Boston, 1968); H. Stafford Hatfield, *The inventor and his world* (New York, 1948).

Technological Dreams

Technological Extrapolations

A. G. Keller, *A theatre of machines* (New York, 1964); Bert S. Hall, "Der Meister sol auch kennen schreiben und lesen: writings about technology ca. 1400–ca. 1600 A.D. and their cultural implications," in *Early technologies*, ed. Denise Schmandt-Besserat (Malibu, Calif., 1979); Agostino Ramelli, *The various and ingenious machines of Agostino Ramelli (1588)*, trans. M. T. Gnudi, ed. E. S. Ferguson (Baltimore, 1976); Eugene S. Ferguson, "The mind's eye: nonverbal thought in technology," *Science* 197 (1977), 827–36.

Patents

Stacy V. Jones, *The patent office* (New York, 1971); IEEE, *Patents and patenting* (New York, 1982); William Ray and Marlys Ray, *The art of invention: patent models and their makers* (Princeton, 1974); Stacy V. Jones, *Inventions necessity is not the mother of* (New York, 1973); A. E. Brown and H. A. Jeffcott, *Absolutely mad inventions* (New York, 1970).

Technological Visions

Gotz Quarg, ed. and trans., *Conrad Kyeser aus Eichstatt, Bellifortis*, 2 vols. (Dusseldorf, 1967); Ladislo Reti, *The unknown Leonardo* (New York, 1974); Charles H. Gibbs-Smith, *The inventions of Leonardo da Vinci* (Oxford, 1978); Martin Kemp, *Leonardo da Vinci* (Cambridge, 1981); Willy Ley, *Engineers' dreams* (New York, 1954); Robert W. Marks, *The dymaxion world of Buckminster Fuller* (New York, 1960); Robert Snyder, ed., *Buckminster Fuller* (New York, 1980); R. Buckminster Fuller, *Operating manual for spaceship earth* (Carbondale, Ill., 1969); Alison Sky and Michelle Stone, *Unbuilt America* (New York, 1976); Frank P. Davidson, L. J. Giacoletto, and Robert Salkeld, eds., *Macro-engineering and the infrastructure of tomorrow* (Boulder, Colo., 1978); Gerard K. O'Neill, *The high frontier: human colonies in space* (New York, 1977); Gerard K. O'Neill, *2081: a hopeful view of the human future* (New York, 1981); T. A. Heppenheimer, *Toward distant suns* (Harrisburg, Penn., 1979); Frank P. Davidson, *Macro: a clear vision of how science and technology will shape our future* (New York, 1983).

Impossible Machines

Arthur W. J. G. Ord-Hume, *Perpetual motion: the history of an obsession* (New York, 1977); Henry Dircks, *Perpetuum mobile*, 1st ser. (London, 1861), 2nd ser. (London, 1870); Theodore Bowie, ed., *The sketchbook of Villard de Honnecourt* (Bloomington, Ind., 1959); Stacey V. Jones, "Motor run solely by magnets," *New York Times*, April 28, 1979, 32; Eliot Marshall, "Newman's impossible motor," *Science* 223 (1984), 571–2.

Popular Fantasies

Lynn Thorndike, *A history of magic and experimental science*, vol. 2 (New York, 1929), pp. 654–5; Thomas P. Dunn and Richard D. Erlich, eds., *The mechanical god, machines in science fiction* (Westport, Conn., 1982); Harry Harrison, *Mechanismo* (Danbury, Conn., 1978); Peter C. Marzio, *Rube Goldberg: his life and work* (New York, 1973); William Heath Robinson, *Inventions*, (London, 1973); William Heath Robinson, *Absurdities*, (London, 1975); John Lewis, *Heath Robinson* (New York, 1973); Jacques Carelman, *Catalog of fantastic things* (New York, 1971); Tim Onosko, *Wasn't the future wonderful?* (New York, 1979); Edward L. Throm, ed., *Fifty years of Popular Mechanics, 1902–1952* (New York, 1952); Ernest V. Heyn, *Fire of genius, inventors of the past century (based on the files of Popular Science Monthly)* (New York, 1976); George Basalla, "Some persistent energy myths," in *Energy and transport: historical perspectives on policy issues*, ed. George H. Daniels and Mark H. Rose (Beverly Hills, Calif., 1982).

Knowledge: Technology Transfer

Imperialism

Noel Perrin, *Giving up the gun: Japan's reversion to the sword, 1543–1879* (Boston, 1979); Lynn White, Jr., *Medieval technology and social change* (New York, 1962); Thomas Francis Carter, *The invention of printing in China and its spread westward*, rev. L. Carrington Goodrich, 2nd ed. (New York, 1955); Daniel R. Headrick, *The tools of empire* (New York, 1981); Percival Spear, *The Oxford history of modern India: 1740–1975*, 2nd ed. (Delhi, 1978); Henry T. Bernstein, *Steamboats on the Ganges* (Bombay, 1960); Michael Satow and Ray Desmond, *Railways of the raj* (New York, 1980).

Migration

Warren C. Scoville, "The Huguenots and the diffusion of technology," *The Journal of Political Economy* 60 (1952), 294–311, 392–411; Carroll W. Pursell, Jr., *Early stationary steam engines in America* (Washington, D.C., 1969); Peter Mathias, "Skills and the diffusion of innovations from Britain in the eighteenth century," *Transactions of the Royal Historical Society*, 5th ser., 25 (1975), 93–113; Eric Robinson, "The early diffusion of steam power," *The Journal of Economic History* 34 (1974), 91–107; Eric Robinson, "The transference of British technology to Russia, 1760–1820: a preliminary enquiry," in *Great Britain and her world, 1750–1914*, ed. Barrie M. Ratcliffe (Manchester, 1975), pp. 1–26; Jennifer Tann and M. J. Breckin, "The international diffusion of the Watt engine, 1775–1825," *The Economic History Review*, 2nd ser., 31 (1978), 541–64; David J. Jeremy, *Transatlantic industrial revolution: the diffusion of textile technologies between Britain and America, 1790–1830s* (Cambridge, Mass., 1981); David J. Jeremy, "British textile technology transmission to the United States: the Philadelphia region experiment, 1770–1820," *Business History Review* 47 (1973), 24–52.

Practical Knowledge

W. H. Chaloner, "Sir Thomas Lombe (1685–1739) and the British silk industry," in *People and industries* (London, 1963); Carlo M. Cipolla, "The diffusion of innovations in early modern Europe," *Comparative Studies in Society and History* 14 (1972), 46–52; Paul Mantoux, *The Industrial Revolution in the eighteenth century*, rev. ed. (New York, 1961); Norman R. Bottom, Jr., and Robert R. J. Gallati, *Industrial espionage* (Boston, 1984); Nick Lyons, *The Sony vision* (New York, 1976); Daniel I. Okimoto, Takuo Sugano, and Franklin B. Weinstein, eds., *Competitive edge: the semiconductor industry in the U.S. and Japan* (Stanford, 1984).

Environmental Influences

Henry J. Kauffman, *American axes* (Brattleboro, Vt., 1972); R. A. Salaman, *Dictionary of tools* (London, 1975); Louis C. Hunter, *Steamboats on western rivers: an economic and technological history* (Cambridge, Mass., 1949); John H. White, Jr., *American locomotives: an engineering history, 1830–1880* (Baltimore, 1968); John F. Stover, *The life and decline of the American railroad* (New York, 1970); Klaus Peter Harder, *Environmental factors of early railroads* (New York, 1981);

James E. Brittain, "The international diffusion of electrical power technology, 1870–1920" (with comments by Thomas P. Hughes), *Journal of Economic History* 34 (1974), 108–28; Thomas P. Hughes, *Networks of power: electrification in Western society: 1810-1930* (Baltimore, 1983).

Knowledge: Science

Hugh G. J. Aitken, *The continuous wave: technology and American radio 1900–1932* (Princeton, 1985); A. E. Musson and Eric Robinson, *Science and technology in the Industrial Revolution* (Manchester, 1969); H. W. Dickinson, *A short history of the steam engine*, introd. A. E. Musson (London, 1963); L. T. C. Rolt, *Thomas Newcomen* (London, 1963); James Patrick Muirhead, *The life of James Watt* (New York, 1859); Eugene S. Ferguson, "The origins of the steam engine," *Scientific American* 210, no. 1 (1964), 98–107; D. S. L. Cardwell, *Steam power in the eighteenth century* (London, 1963); Eugene S. Ferguson, "The mind's eye: nonverbal thought in technology," *Science* 197 (1977), 827–36; Hugh G. J. Aitken, *Syntony and spark – the origins of radio* (New York, 1976); Morris Kline, *Mathematics and the physical world* (New York, 1959); James Clerk Maxwell, *The scientific papers of James Clerk Maxwell*, ed. W. D. Niven, vol. 2 (Cambridge, 1890); W. J. Jolly, *Sir Oliver Lodge* (Rutherford, 1974).

IV. NOVELTY (2): SOCIOECONOMIC AND CULTURAL FACTORS

Making Things by Hand

H. G. Barnett, *Innovation: the basis of cultural change* (New York, 1953); Robert S. Merrill, "Routine Innovation" (Ph.D. diss., The University of Chicago, 1959); David Pye, *The nature and art of workmanship* (Cambridge, 1968); Lila M. O'Neale, *Yurok-Karok basket weavers* (Berkeley, Calif., 1932); Ruth L. Bunzel, *The Pueblo potter: a study of creative imagination in primitive art* (New York, 1929); May N. Diaz, *Tonalá; conservatism, responsibility and authority in a Mexican town* (Berkeley, Calif., 1966); George Caspar Homans, *Social behavior: its elementary forms* (New York, 1974); Dean E. Arnold, *Ceramic theory and cultural process* (Cambridge, 1985); Harry R. Silver, "Calculating risks: the socioeconomic foundations of aesthetic innovation in an Ashanti carving community," *Ethnology* 20 (1981), 101–14; Neil Cossons and Barrie Trinder, *The iron bridge* (Bradford-on-Avon, 1979); Eric S. de Maré, *The bridges of Britain* (London,

1954); R. U. Sayce, *Primitive arts and crafts* (Cambridge, 1933); Philip Steadman, *The evolution of designs* (Cambridge, 1979); D. S. Robertson, *Greek & Roman architecture* (Cambridge, Mass., 1969); Christopher Alexander, *Notes on the synthesis of form* (Cambridge, 1964); Theodore Wertime, "Pyrotechnology: man's fire-using crafts" in *Early technologies*, ed. Denise Schmandt-Besserat (Malibu, 1979).

Economic Incentives

Nathan Rosenberg, *Perspectives on technology* (Cambridge, 1976); Nathan Rosenberg, *Inside the black box: technology and economics* (Cambridge, 1982); Nathan Rosenberg, ed., *The economics of technological change* (Harmondsworth, 1971); Karl Marx, "Manifesto of the Communist Party," in *On revolution*, ed. Saul K. Padover, vol. 1 (New York, 1972), pp. 79–107; Karl Marx, *Capital*, vol. 1, trans. Samuel Moore and Edward Aveling (New York, 1967); Tine Bruland, "Industrial conflict as a source of technical innovation: three cases," *Economy and Society* 11 (1982), 91–121; G. N. Clark, *Science and social welfare in the age of Newton* (London, 1937).

Market Demand

Joseph A. Schumpeter, *Business cycles*, vol. 1 (New York, 1939); Jacob Schmookler, *Invention and Economic Growth* (Cambridge, Mass., 1966).

Labor Scarcity

Nathan Rosenberg, ed., *The American system of manufactures* (Edinburgh, 1969); H. J. Habakkuk, *American and British technology in the nineteenth century: the search for labour-saving inventions* (Cambridge, 1962); S. B. Saul, ed., *Technological change: the United States and Britain in the nineteenth century* (London, 1970); Eugene S. Ferguson, *Oliver Evans: inventive genius of the American Industrial Revolution* (Greenville, 1980); Merritt Roe Smith, *Harper's Ferry armory and the new technology* (Ithaca, 1977); David A. Hounshell, *From the American system to mass production, 1800–1932* (Baltimore, 1984); David A. Hounshell, "The system: theory and practice," in *Yankee enterprise: the rise of the American system of manufactures*, ed. Otto Mayr and Robert C. Post (Washington, D.C., 1981), pp. 127–52; Paul Uselding, "Studies of technology in economic history," in *Recent developments in the study of business and economic history: essays in memory of Herman E. Krooss*, ed. Robert E. Gallman (Greenwich, Conn., 1977), pp. 159–219.

of Herman E. Krooss, ed. Robert E. Gallman (Greenwich, Conn., 1977), pp. 159–219.

Patents

Rupert T. Gould, *The marine chronometer* (London, 1960); Humphrey Quill, *John Harrison: the man who found longitude* (New York, 1966); Derek Hudson and Kenneth W. Luckhurst, *The Royal Society of Arts, 1754–1954* (London, 1954); Edgar Burke Inlow, *The patent grant* (Baltimore, 1950); Bruce Willis Bugbee, *Genesis of American patent and copyright law* (Washington, D.C., 1967); Morgan Sherwood, "The origins and development of the American patent system," *American Scientist* 71 (1983), 500–6; Nathan Rosenberg, *Technology and American economic growth* (New York, 1972); Gerhard Rosegger, *The economics of production and innovation* (Oxford, 1980); H. I. Dutton, *The patent system and inventive activity during the Industrial Revolution: 1750–1852* (Manchester, 1984); Fritz Machlup, "Patents and inventive effort," *Science* 133 (1961), 1463–6; David F. Noble, *America by design: science, technology and the rise of corporate capitalism* (New York, 1977); C. T. Taylor and Z. A. Silberston, *The economic impact of the patent system: a study of the British experience* (Cambridge, 1973); Eric Schiff, *Industrialization without national patents* (Princeton, 1971); *Great Soviet encyclopedia*, 3rd ed., s.v. "Author's certificate," by I. A. Gringol.

Industrial Research Laboratories

J. D. Bernal, *Science and industry in the nineteenth century* (Bloomington, Ind., 1970); Georg Meyer-Thurow, "The industrialization of invention: a case study from the German chemical industry," *ISIS* 73 (1982), 363–81; John Joseph Beer, *The emergence of the German dye industry* (Urbana, Ill., 1959); Kendall A. Birr, "Science in American industry," in *Science and society in the United States*, ed. David D. Van Tassel and Michael G. Hall (Homewood, Ill., 1966); Kendall Birr, *Pioneering in industrial research: the story of the General Electric Research Laboratory* (Washington, D. C., 1957); George Wise, "A new role for professional scientists in industry: industrial research at General Electric, 1900–1916," *Technology and Culture* 21 (1980), 408–29; Lillian Hoddeson, "The Emergence of Basic Research in the Bell Telephone System, 1875–1915," *Technology and Culture* 22 (1981), 512–44; Leonard S. Reich, *The making of American industrial research* (New York, 1985); Leonard S. Reich, "Irving Langmuir and the pursuit of science and technology in the corporate environment," *Technology and Culture* 24 (1983), 199–221; Leonard S. Reich, "Science," in *Encyclopedia of American economic history*, vol. 1, ed. Glenn Porter (New York, 1980); Leonard S. Reich, "Industrial research and the pursuit of corporate security: the early years of Bell Labs," *Business History Review* 54 (1980), 504–29; Leonard S. Reich, "Research, patents, and the struggle to control radio: a study of the uses of industrial research," *Business History Review* 51 (1977), 208–35; Willard F. Mueller, "The origins of the basic inventions underlying Du Pont's major product and process innovations, 1920–1950," in National Bureau of Economic Research, *The rate and direction of inventive activity: economic and social factors* (Princeton, 1962); John Jewkes, David Sawers, Richard Stillerman, *The sources of invention*, 2nd ed. (London, 1969).

Novelty and Culture

G. N. Clark, *Science and social welfare in the age of Newton* (London, 1949); Denys Hay, *Polydore Vergil* (London, 1952); Francis Bacon, *The great instauration and new Atlantis*, ed. J. Weinberger (Arlington Heights, Ill., 1980); Paolo Rossi, *Francis Bacon: from magic to science* (London, 1968); Benjamin Farrington, *Francis Bacon, philosopher of industrial science* (New York, 1949); J. G. Crowther, *Francis Bacon* (London, 1960); Bernard Lewis, *The Muslim discovery of Europe* (New York, 1982); Lynn Thorndike, "Newness and craving for novelty in seventeenth-century science and medicine," *The Journal of the History of Ideas* 12 (1951), 584–98; J. B. Bury, *The idea of progress* (New York, 1932); Leslie Sklair, *The sociology of progress* (London, 1970); Paolo Rossi, *Philosophy, technology and the arts in the early modern era* (New York, 1970); Richard F. Jones, *Ancients and moderns: a study of the rise of the scientific movement in seventeenth-century England* (St. Louis, 1961); Clarence J. Glacken, *Traces on the Rhodian shore: nature and culture in Western thought* (Berkeley, Calif., 1967); William Leiss, *The domination of nature* (New York, 1972); Lynn White, Jr., "The historical roots of our ecologic crisis," *Science* 155 (1967), 1203–7.

V. SELECTION (1): ECONOMIC AND MILITARY FACTORS

Introduction

Philip Steadman, *The evolution of designs* (London, 1979); Alfred L. Kroeber, *Anthropology* (New York, 1948).

General Considerations

Roland Gelatt, *The fabulous phonograph, 1877–1977*, 2nd ed. (New York, 1977); Robert E. Conot, *A streak of luck* (New York, 1979); Nick Lyons, *The Sony vision*, (New York, 1976).

Economic Constraints

Nathan Rosenberg, "The influence of market demand upon innovation," in *Inside the black box: technology and economics* (London, 1982).

The Waterwheel and the Steam Engine

Terry S. Reynolds, *Stronger than a hundred men: a history of the vertical water wheel* (Baltimore, 1983); L. A. Moritz, *Grain-mills and flour in classical antiquity* (Oxford, 1958); Robert H. J. Sellin, "The large Roman water mill at Barbegal France," in *History of technology – eighth annual volume, 1983* ed. Norman Smith (London, 1983); Lynn White Jr., *Medieval technology and social change* (London, 1962); Jean Gimpel, *The medieval machine: the Industrial Revolution in the Middle Ages* (New York, 1976); Marc Bloch, "The advent and triumph of the watermill," in *Land and work in the Medieval Europe* (London, 1967); Louis C. Hunter, *A history of industrial power in the United States, 1780–1930*, vol. 1: *waterpower in the century of the steam engine* (Charlottesville, N.C., 1979); Louis C. Hunter, "The living past in the Appalachias of Europe: water-mills in southern Europe," *Technology and Culture* 8 (1967), 446–66; G. N. von Tunzelman, *Steam power and British industrialization to 1860* (London, 1978).

The Mechanical Reaper

Michael Partridge, *Farm tools through the ages* (Reading, 1973); William T. Hutchinson, *Cyrus Hall McCormick: seed-time, 1809–1856*, vol. 1 (New York, 1930); Cyrus McCormick, *The century of the reaper* (Boston, 1931); C. H. Wendel, *150 years of International Harvester* (Sarasota, Fla., 1981); Siegfried Giedion, *Mechanization takes command* (New York, 1969), pp. 146–62; Paul A. David, "The mechanization of reaping in the ante-bellum Midwest," in *Technical choice, innovation, and economic growth*, ed. Paul A. David (London, 1975), pp. 195–232; L. J. Jones, "The early history of mechanical harvesting," in *History of Technology*, vol. 4, ed. A. Rupert Hall and Norman Smith (London, 1979), pp. 101–48; Merrill Denison, *Harvest triumphant: the story of Massey-Harris* (New York, 1949), p. 12.

Merrill Denison, *Harvest triumphant: the story of Massey-Harris* (New York, 1949), p. 12.

The Supersonic Transport

Mel Horwitch, *Clipped wings: the American SST conflict* (Cambridge, Mass., 1982)

Military Necessity

Merritt Roe Smith, ed., *Military enterprise and technological change* (Cambridge, Mass., 1985); William H. McNeill, *The pursuit of power: technology, armed force, and society since A.D. 1000* (Chicago, 1982); John U. Nef, *War and human progress: an essay on the rise of industrial civilization* (Cambridge, 1950); J. M. Winter, ed., *War and economic development* (Cambridge, 1975); Michael S. Sherry, *Preparing for the next war: American plans for postwar defense, 1941–45* (New Haven, Conn., 1977).

The Motor Truck

James J. Flink, *America adopts the automobile, 1895–1910* (Cambridge, Mass., 1970); James J. Flink, *The car culture* (Cambridge, Mass., 1975); Robert F. Karolevitz, *This was trucking: a pictorial history of the first quarter century of commercial motor vehicles* (Seattle, 1966); James A. Wren and Genevieve J. Wren, *Motor trucks of America* (Ann Arbor, 1979); James A. Huston, *The sinews of war: army logistics, 1775–1953* (Washington, D.C., 1966); U.S. Department of Transportation, *America's highways: 1776–1976* (Washington, D.C., 1977).

Nuclear Power

Frederick Soddy, *The interpretation of radium* (New York, 1912); Thaddeus J. Trenn, *The self-splitting atom: the history of the Rutherford–Soddy collaboration* (London, 1977); H. G. Wells, *The world set free*, repr. ed. (London, 1956); Stephen Hilgartner, Richard C. Bell, and Rory O'Connor, *Nukespeak: nuclear language, visions, and mindset* (San Francisco, 1982); Spencer R. Weart and Gertrud W. Szilard, eds., *Leo Szilard: his versions of the facts* (Cambridge, Mass., 1978); Hans G. Graetzer and David L. Anderson, *The discovery of nuclear fission: a documentary history* (New York, 1971); Richard G. Hewlett and Oscar E. Anderson, Jr., *The new world, 1939/1946, Vol. l: a history of The United States Atomic Energy Commission* (University Park, Penn., 1962); Gerard H. Clarfield and William M. Wiecek, *Nuclear America: military and civilian nuclear power in the United States, 1940–1980* (New York, 1984); Peter Pringle and James Spigelman, *The nuclear barons* (New York, 1981); Richard G. Hewlett and Francis Duncan, *Nuclear navy, 1946–1962* (Chicago, 1974); Irvin C. Bupp and Jean-Claude Derian, *Light water: how the dream dissolved* (New York, 1978); Walter C. Patterson, *Nuclear power*, 2nd ed. (London, 1983); Robert Perry et al., *Development and commercialization of the light water reactor, 1946–1976* (R–2180–NSF), Rand Corporation (Santa Monica, Calif., 1977); Wendy Allen, *Nuclear reactors for generating electricity: U.S. development from 1946 to 1963* (R–2116–NSF), Rand Corporation (Santa Monica, Calif., 1977); Daniel J. Kevles, *The physicists: the history of a scientific community in modern America* (New York, 1978).

VI. SELECTION (2): SOCIAL AND CULTURAL FACTORS

Technology and Chinese Culture

Joseph Needham, with Wang Ling, *Science and civilization in China*, 6 vols. (Cambridge, 1954–84); Joseph Needham, *The grand titration, science and society in East and West* (Toronto, 1969); Joseph Needham, *Science in traditional China* (Cambridge, 1981).

Printing

Tsuen-hsuin Tsien, *Written on bamboo and silk: the beginnings of Chinese books and inscriptions* (Chicago, 1962); Thomas Francis Carter, *The invention of printing in China and its spread westward*, 2nd ed. rev. by L. Carrington Goodrich (New York, 1955); Colin Clair, *A history of European printing* (London, 1976); S. H. Steinberg, *Five hundred years of printing* (London, 1959); Elizabeth L. Eisenstein, *The printing press as an agent of change*, 2 vols. (Cambridge, 1979); A. Stevenson, "The quincentennial of Netherlandish block books," *The British Museum Quarterly* 31 (1965), 85–9.

Gunpowder

Nathan Sivin, ed., *Science and technology in East Asia* (New York, 1977); Carlo M. Cipolla, *Guns, sails, and empires* (New York, 1965); William H. McNeill, *The pursuit of power* (Chicago, 1982); Carlo M. Cipolla, *Clocks and culture, 1300–1700* (New York, 1967).

The Magnetic Compass

E. G. R. Taylor, *The haven-finding art* (New York, 1971); J. H. Parry, *The age of reconnaissance* (Berkeley, Calif., 1981).

Chinese Cultural Stagnation

Nathan Sivin and Shigeru Nakayama, eds., *Chinese science* (Cambridge, Mass., 1973); Nathan Sivin, "Why the scientific revolution did not take place in China − or didn't it?" in *Transformation and tradition in the sciences*, ed. Everett Mendelsohn (Cambridge, 1984); Mark Elvin, *The pattern of the Chinese past* (Stanford, Calif., 1973); Joseph R. Levenson, *Confucian China and its modern fate* (Berkeley, Calif., 1958).

Fads and Fashions

Atmospheric Railway

Hamilton Ellis, *British railway history (1830–1876)* (London, 1954); Charles Hadfield, *Atmospheric railways: a Victorian venture in silent speed* (New York, 1968); L. T. C. Rolt, *Isambard Kingdom Brunel: a biography* (London, 1959); Peter R. Hay, *Brunel: his achievements in the transport revolution* (Reading, 1973); Sir Alfred Pugsley, ed., *The works of Isambard Kingdom Brunel* (London, 1976); Derrick Beckett, *Brunel's Britain* (Newton Abbot, 1980).

Nuclear Propulsion Vehicles

Joseph J. Corn, *The winged gospel: America's romance with aviation, 1900–1950* (New York, 1983); Peter Pringle and James Spigelman, *The nuclear barons* (New York, 1981); H. Peter Metzger, *The atomic establishment* (New York, 1972); Stephen Hilgartner, Richard C. Bell, and Rory O'Connor, *Nukespeak* (San Francisco, 1982); Freeman J. Dyson, "Death of a project," *Science* 149 (1965), 141–4; Freeman J. Dyson, *Disturbing the universe* (New York, 1979); Herbert York, *Race to oblivion* (New York, 1970); W. Henry Lambright, *Shooting down the nuclear plane* (Syracuse, N.Y., 1967); Walter C. Patterson, *Nuclear power* (Harmondsworth, 1983); A. W. Kramer, *Nuclear propulsion for merchant ships* (Washington, D.C., 1962); David Kuechle, *The story of the Savannah* (Cambridge, Mass., 1971); "Whatever happened to America's atom-powered merchant ship," *U.S. News and World Report*, August 16, 1971, 49; David E. Sanger, "The expected boom in home computers fails to materialize," *New York Times*, June 4, 1984, p. 1.

"The expected boom in home computers fails to materialize," *New York Times*, June 4, 1984, p. 1.

Discard and Extinction

W. H. R. Rivers, "The disappearance of useful arts," in *Psychology and ethnology* (New York, 1926); George Kubler, *The shape of time: remarks on the history of things* (New Haven, Conn., 1962); Matthew H. Nitecki, ed., *Extinctions* (Chicago, 1984); Christopher S. Wren, "In China, the steam engine is still king of the rails," *New York Times*, October 10, 1984, p. 2; Noel Perrin, *Giving up the gun: Japan's reversion to the sword, 1543–1879* (Boston, 1979).

Alternative Paths

Hand Tools

Paul B. Kebabian, *American woodworking tools* (Boston, 1978); Robert F. G. Spier, *From the hand of man* (Boston, 1970); Toshio Odate, *Japanese woodworking tools* (Newtown, 1984).

Block Printing: East and West

Marshall McLuhan, *The Gutenberg galaxy* (Toronto, 1962); David S. Landes, *Revolution in time: clocks and the making of the modern world* (Cambridge, Mass., 1983); D. S. L. Cardwell, *Turning points in Western technology* (New York, 1972); Thomas Francis Carter, *The invention of printing in China and its spread westward*, 2nd ed., rev. by L. Carrington Goodrich (New York, 1955); Tsuen-hsuin Tsien, *Written on bamboo and silk: the beginnings of Chinese books and inscriptions* (Chicago, 1962); Maurice Audin, "Printing," in *A history of technology and invention*, vol. 2, ed. Maurice Daumas (New York, 1969); Jonathan D. Spence, *The memory palace of Matteo Ricci* (New York, 1984); Hendrik D. L. Vervliet, *The book through five thousand years* (London, 1972); Colin Clair, *A history of European printing* (London, 1976); Joseph Needham, *Science and civilization in China*, vol. 5, pt. 1: *Paper and printing* (Cambridge, 1985); A. Stevenson, "The quincentennial of Netherlandish block books," *The British Museum Quarterly* 31 (1965), 85–9.

Railroads versus Canals

Robert William Fogel, *Railroads and American economic growth* (Baltimore, 1964); Robert William Fogel, "Railroads as an analogy to the space effort: some economic aspects," in *The railroad and the space program*, ed. Bruce Mazlish (Cambridge, Mass., 1965); Robert William Fogel, "Notes on the social saving controversy," *The Journal of Economic History* 39 (1979), 1–54; Peter D. McClelland, "Transportation," in *Encyclopedia of American economic history*, vol. 1, ed. Glenn Porter (New York, 1980), 309–34; Francis T. Evans, "Roads, railways, and canals: technical choices in 19th-century Britain," *Technology and Culture* 22 (1981), 1–34.

Steam, Electric, and Gasoline Vehicles

James J. Flink, *America adopts the automobile: 1895–1910* (Cambridge, Mass., 1970); James J. Flink, *The car culture* (Cambridge, Mass., 1975); John B. Rae, *The American automobile: a brief history* (Chicago, 1965); Richard H. Schallenberg, *Bottled energy: electrical engineering and the evolution of chemical energy storage* (Philadelphia, 1982); Allan Nevins, *Ford: the times, the man, the company*, vol. 1 (New York, 1954); Robert F. Karolevits, *This was trucking* (Seattle, Wash., 1966); Edward E. La Schum, *The electric motor truck* (New York, 1924); Charles C. McLaughlin, "The Stanley Steamer: a study in unsuccessful innovation," *Explorations in Entrepreneurial History* 7 (1954), 37–47; John B. Rae, "The engineer–entrepreneur in the American automobile industry," *Explorations in Entrepreneurial History* 8 (1955), 1–11; John Bentley, *Oldtime steam cars* (New York, 1969); Andrew Jamison, *The steam-powered automobile: an answer to air pollution* (Bloomington, Ind., 1970); Society of Automotive Engineers, *Energy and the automobile* (New York, 1973); Stuart W. Leslie, *Boss Kettering* (New York, 1983); Nathan Rosenberg, *Inside the black box: technology and economics* (Cambridge, 1982).

Conclusion

Langdon Winner, *Autonomous technology* (Cambridge, Mass., 1977).

VII. CONCLUSION: EVOLUTION AND PROGRESS

Evolution

Langdon Winner, *Autonomous technology* (Cambridge, Mass., 1977); Karl Marx, *Capital*, vol. 1, trans. Samuel Moore and Edward Aveling (New York, 1967).

Technological Progress

J. B. Bury, *The idea of progress* (New York, 1932); Robert Nisbet, *History of the idea of progress* (New York, 1980); Leslie Sklair, *The sociology of progress* (London, 1970); Gabriel A. Almond, Marvin Chodorow, and Roy H. Pearce, eds., *Progress and its discontents* (Berkeley, 1982); Donella H. Meadows, Dennis L. Meadows, Jørgen Randers, and William W. Behrens III, *The limits to growth* (New York, 1972); John Maddox, *The doomsday syndrome* (New York, 1972); David Pimentel and Marcia Pimentel, *Food, energy and society* (New York, 1979); Sally Green, *Prehistorian: a biography of V. Gordon Childe* (Bradford-on-Avon, 1981); V. Gordon Childe, *Man makes himself* (New York, 1951); V. Gordon Childe, *What happened in history* (Baltimore, 1952); V. Gordon Childe, *Progress and archaeology* (London, 1944); V. Gordon Childe, *Social evolution* (New York, 1951); V. Gordon Childe, *Society and knowledge* (London, 1956); Phyllis Deane, *The first industrial revolution* (Cambridge, 1979).

Sources of Quotations

I. DIVERSITY, NECESSITY, EVOLUTION

1 Thomas J. Schlereth, *Material culture studies in America* (Nashville, 1982), p. 2.
2 e. e. cummings, "pity this busy monster manunkind," in *Poems, 1923–1954* (New York, 1954), p. 397.
3 David S. Landes, *Revolution in time: clocks and the making of the modern world* (Cambridge, Mass., 1983), p. 6.
4 S. C. Gilfillan, *The sociology of invention* (1935; repr. Cambridge, Mass., 1970), p. 24.

II. CONTINUITY AND DISCONTINUITY

1 Edward W. Constant, *The origins of the turbojet revolution* (Baltimore, 1980), p. 19.
2 Cooke-Taylor, quoted in E. P. Thompson, *The making of the English working class* (New York, 1960), p. 190.
3 Joseph Needham, *Clerks and craftsmen in China and the West* (Cambridge, 1970), p. 202.
4 Ernest Braun & Stuart Macdonald, *Revolution in miniature* (Cambridge, 1982), p. 1.
5 "The transistor," *Bell Laboratories Record* 26 (1948), 322.
6 Thomas A. Edison, quoted in Harold C. Passer, "The electric light and gas light: innovation and continuity in economic history," *Explorations in Entrepreneurial History* 1 (1949), 2.
7 *Ibid.*, p. 3.
8 Jesse S. James, *Early United States barbed wire patents* (Maywood, Ill., 1966), p. 3.
9 Henry D. and Frances T. McCallum, *The wire that fenced the West* (Norman, Okla., 1965), p. 23.
10 John Locke, *Some thoughts concerning education* (London, 1699), pp. 272–3.
11 John C. Kimball, quoted in John F. Kasson, *Civilizing the machine* (New York, 1976), p. 153.

241

12 Friedrich Engels, *The condition of the working class in England,* trans. W. O. Henderson and W. H. Chaloner (Oxford, 1971), p. 9.

III. NOVELTY (1): PSYCHOLOGICAL AND INTELLECTUAL FACTORS

1 Eugene S. Ferguson, "The mind's eye: nonverbal thought in technology," *Science* 197 (1977), 829.
2 Jennifer Tann and M. J. Breckin, "The international diffusion of the Watt engine," *The Economic History Review* 31 (1978), 557.
3 David J. Jeremy, "British textile technology transmission to the United States: the Philadelphia region experience, 1770–1820," *Business History Review* 47 (1973), 26.
4 Carlo M. Cipolla, "The diffusion of innovations in early modern Europe," *Comparative Studies in Society and History* 14 (1972), 47.
5 "Recueil de diverses pieces touchant quelques nouvelles machines, par M. D. Papin," *Philosophical Transactions of the Royal Society of London* 19 (1697), 482.
6 James Clerk Maxwell, *The scientific papers of James Clerk Maxwell,* ed. W. D. Niven, vol. 2 (Cambridge, 1890), p. 742.
7 Hugh G. J. Aitken, *Syntony and spark – the origins of radio* (New York, 1976), pp. 204–5.
8 Friedrich Kurylo and Charles Susskind, *Ferdinand Braun* (Cambridge, Mass., 1981), p. 226.

IV. NOVELTY (2): SOCIOECONOMIC AND CULTURAL FACTORS

1 H. G. Barnett, *Innovation: the basis of cultural change* (New York, 1953), p. 49.
2 Karl Marx, "Manifesto of the Communist Party," in *On revolution,* ed. Saul K. Padover, vol. 1 (New York, 1972), p. 83.
3 Karl Marx, *Capital,* vol. 1, trans. Samuel Moore and Edward Aveling (New York, 1967), p. 436.
4 H. J. Habakkuk, *American and British technology in the nineteenth century* (Cambridge, 1967), p. 99.
5 Morgan Sherwood, "The origins and development of the American patent system," *American Scientist* 71 (1983), 501.
6 Fritz Machlup, quoted in Gerhard Rosegger, *The economics of production and innovation* (Oxford, 1980), p. 190.
7 *Great Soviet encyclopedia,* 3rd. ed., s.v. "Author's certificate," by I. A. Gringol.
8 Matthew Josephson, *Edison* (New York, 1959), pp. 133–4.
9 E. W. Rice, quoted in Kendall Birr, *Pioneering in industrial research* (Washington, D.C., 1957), p. 31.
10 Leonard S. Reich, "Research, patents, and the struggle to control radio," *Business History Review* 51 (1977), 231.
11 Willard F. Mueller, "The origins of the basic inventions underlying Du Pont's major product and process innovations, 1920–1950," in National Bureau of Economic Research, *The rate and direction of inventive activity: economic and social factors* (Princeton, 1962), p. 323.
12 Bernard Lewis, *The Muslim discovery of Europe* (New York, 1982), p. 224.

13 Lynn Thorndike, "Newness and craving for novelty in seventeenth-century science and medicine," *Journal of the History of Ideas* 12 (1951), 598.
14 Francis Bacon, *Novum organum*, book 2, aphorism 31.

V. SELECTION (1): ECONOMIC AND MILITARY FACTORS

1 Robert Conot, *A streak of luck* (New York, 1979), p. 245.
2 Robert F. Karolevitz, *This was trucking* (Seattle, 1966), p. 65.
3 Ibid.
4 Gerard H. Clarfield and William M. Wiecek, *Nuclear America* (New York, 1984), p. 22.

VI. SELECTION (2): SOCIAL AND CULTURAL FACTORS

1 Charles Hadfield, *Atmospheric railways* (New York, 1968), p. 73.
2 Freeman Dyson, *Disturbing the universe* (New York, 1979), p. 115.
3 Robert Fogel, *Railroads and American economic growth* (Baltimore, 1964), p. 8.
4 Langdon Winner, *Autonomous technology* (Cambridge, Mass., 1977), p. 296.

VII. CONCLUSION: EVOLUTION AND PROGRESS

1 Karl Marx, *Capital*, vol. 1, trans. Samuel Moore and Edward Aveling (New York, 1967), p. 372.
2 V. Gordon Childe, *Social evolution* (New York, 1951), p. 9.
3 V. Gordon Childe, *Man makes himself* (New York, 1951), p. 186.
4 Ibid., p. 19.

Index

Aesop, 3, 6, 13
agriculture, 14, 116; modern, 212–13; primitive, 212–13
airplane, 142, 181–2, 217
Aitken, Hugh G. J., 99
Alexander, Christopher, 108
alphabet blocks, 57
alternative paths, 189–204
American system of manufacturing, 116, 118–19
analogies, organic–mechanical, 14–21
animal kingdom, 13, 14
animal tool use, 13
Aristotle, 15
Arkwright, Richard, 28
artifact, 30; hand-crafted, 103–10; mass-produced, 103–4
Atoms for Peace, 163, 165, 184
Australian aborigines, 18, 64
automobile, 6, 7, 91, 138, 143, 159, 212; electric, 198–200, 201, 202; gasoline, 198, 200, 201–3; steam, 197, 198, 200–2, 203
axe, 88–9, 108

Bachelard, Gaston, 14
Bacon, Francis, 129, 130, 132, 133, 169, 172, 176, 192
Bacon, Roger, 75–6
Bardeen, John, 44
Barnett, H. G., 103
baskets, 104, 106
Bell, Alexander Graham, 98, 141
Bell laboratories, 44, 45, 126, 127
beverage can, 103–4
bicycle, 57
book-writing machine, 55–7
Boulton, Matthew, 82

bow and arrow, 186
Brattain, Walter H., 44
Braun, Ferdinand, 44, 101
Brittain, James E., 91
Bruland, Tine, 111
Brunel, Isambard K., 177, 179–80, 181
Bunzel, Ruth L., 104–5
Butler, Samuel, 15, 16, 24

calculators, mechanical, 56
camel, 10, 11
camera, 142, 217
canals, 195, 196, 197, 203
canoe, 108, 186
Capek, Karel, 76
Cardwell, D. S. L., 50
Carelman, Jacques, 76
change, cumulative, 21–4
chemistry, organic, 28, 124–5
Childe, V. Gordon, 212, 213–16
classification, 16, 17, 137
clock, mechanical, 7, 56, 59
Coalbrookdale, England, 106
compass, magnetic, 41, 169, 172, 173, 176
computer, 141, 143, 185
Constant, Edward W., 28–9
continuity (case studies): barbed wire, 49–55; book-writing machine, 55–7; cotton gin, 32–4; Edison's lighting system, 46–9; electric motor, 40–3; steam and internal combustion engines, 35–40; stone tools, 30–2; transistor, 43–6
continuity, concept of, 21, 26, 208–9
Cooke-Taylor, W., 35
cooking, 14
cotton gin: Whitney, 32–4, 35, 62; *charka*, 33–4, 57, 62
Crochett, Thomas, 84

Crompton, Samuel, 28
Crookes, Sir William, 99
cummings, e. e., 2

Darwin, Charles, 1, 21, 207–8, 217
Daumas, Maurice, 40
David, Paul A., 153
De Forest, Lee, 44
design process, 108
detector, crystal, 44, 45
Diaz, May N., 104–5
Diderot, Denis, 33
Dillon, Sidney, 195
dirigibles, 188
discard, 185–8
discontinuity, 21, 26, 57–62
diversity, 1, 2, 64, 108, 187, 208
dreams, technological, 67–73
Du Pont Company, 125, 128
duration, 187, 189; intermittent, 188, 189
Dyson, Freeman J., 182–3

Eastman Kodak, 126
Edison, Thomas A., 46–9, 60, 61, 125, 139–40, 141–2, 143, 217
Einstein, Albert, 162
Eisenhower, Dwight D., 163, 164
electricity, 28, 125
electromagnetic waves, 97–102, 216–17
electromagnetism, discovery of, 41
Ellwood, Isaac L., 53
Engels, Friedrich, 61
Ericsson, John, 40
Evans, Oliver, 117
evolution: organic, 1, 2, 3, 15, 16, 20, 135–9, 207, 213–14, 217–18; technological, 1, 2, 3, 15, 16, 30, 61, 135–9, 207–10, 213–15, 217–18
extinction, 185–7, 188, 189
extrapolations, technological, 67–9

fantasies, popular technological, 74–7
Faraday, Michael, 41, 97, 101
Federal Aviation Agency (FAA), 155, 156, 157, 158
fence: barbed wire, 50–5; hedge row, 51–4; smooth wire, 52; stone or wood, 51
Ferguson, Eugene S., 69, 97
Fermi, Enrico, 162
fire, 7, 11, 13
firearms: in Japan, 78–9, 188–9
Firth, Raymond, 65–6
Fleming, John A., 44
Flink, James J., 159
flour mill, automatic, 117
Fogel, Robert W., 195–7
Ford, Henry, 57, 143, 181, 198
Fulton, Robert, 151

Galilei, Galileo, 92, 130
General Electric Company, 125–6
generator, electric power, 91, 112, 165, 166, 167
Gilfillan, S. C., 21–3, 24
Gille, Paul, 40
Glidden, Joseph F., 53
Goldberg, Rube, 76
grain mills, 144–6, 147, 148
Great Exhibition of 1851, 59, 115–16
Greenwalt, Crawford H., 128
Groves, General Leslie R., 162
Guericke, Otto von, 92
gunpowder, 169, 171–2, 173, 176
Gutenberg, Johann, 192, 194

Habakkuk, H. J., 116–19
Haish, Jacob, 53, 54
hammers, 2
Hargreaves, James, 28
Henry, Joseph, 41, 61
Hertz, Heinrich, 98–9, 100, 101, 102, 216, 217
Hicks, John R., 115
Hindle, Brooke, 30
Homans, George C., 105
homo faber, 66
homo ludens, 66
Hornblower, Josiah, 82
Horwitch, Mel, 156
hot air engine, 40
Hughes, Thomas P., 91
Huguenots, 81, 84
Huygens, Christiaan, 92

industrial espionage, 86
industrial research laboratories, 124–7
Industrial Revolution, 27–8, 59, 61–2, 113, 122, 214, 215
innovation, routine, 104, 108
Insull, Samuel, 141
integrated circuit, 45–6
internal combustion engine, 40, 142, 197, 198, 200–3, 217
invention: cumulative synthesis approach to, 23; heroic theory of, 21, 26, 59–60; potential for, 65, 134; psychological aspects of, 24, 64–5, 66; as social process, 21, 103
inventions: capital goods, 114, 115; labor-saving, 115–19; and play, 66, 67; selection of, 139–43
inventor: corporate, 121; as entrepreneur, 151; heroic, 26, 34, 59–60; independent, 128; modern, 130; Renaissance, 129; in U.S.S.R., 123–4

Jeremy, David J., 83
Jervis, John B., 90–1

Johnson, Lyndon B., 156
journalism, scientific/technical, 77

Kay, John, 28
Kelly, Michael, 53
Kennedy, John F., 155
Kepler, Johannes, 130
knowledge: scientific, 91–102; technological, 78–91
Kroeber, Alfred L., 137–8
Kubler, George, 187, 188
Kuhn, Thomas S., 28
Kyeser, Conrad, 71

Landes, David S., 7, 192
language, 12
Laplace, Pierre-Simon de, 209
Lenoir, Jean Joseph Etienne, 40
Leonardo da Vinci, 71–2
light bulb, incandescent, 48, 217
lighting systems, 46–9
Locke, John, 57
locomotive, steam, 90–1, 188
Lodge, Sir Oliver, 98–9, 100, 101, 102, 216
Lodygin, A. N., 60
Lombe, John, 84
Lombe, Sir Thomas, 84

McCormick, Cyrus H., 63, 116, 151–4
machine tools, 16, 118
machines: imaginary, 55; impossible, 73–4; fantastic, 57; perpetual motion, 73–4
McLuhan, Marshall, 192
Marconi, Guglielmo, 60, 100–2, 216–17
Martinez, Julian and Maria, 105
Marx, Karl, 2, 13, 21, 81, 110, 207–8
materials, new, 106
Maxwell, James Clerk, 97–8, 99, 100, 101, 102, 216
Mesopotamia, 8, 9
metaphor, 2, 3
Morse, Samuel F. B., 60–1, 80, 151
motor, electric, 40–3
motor truck, 7, 159–61
Mueller, W. F., 128

nature, 4; conquest of, 132–3
naturfact, 50, 55
necessity, 2, 3–7, 208, 218
Needham, Joseph, 40, 174–5
needs: fundamental, 12–14, 66, 218; perceived, 14, 218
Neolithic, 31
Newcomen, Thomas, 35, 37, 40, 92, 93–7, 102
Nixon, Richard M., 56
novelty, 25, 63, 134, 209–10; artifactual diversity, 64, 65; culture, 64, 65, 129–

33; economic incentives, 110–19; excess of, 135; fantasy, 66–78; hand-crafted artifact, 103–10; industrial conflict, 110–12; industrial research laboratories, 124–9; Islam, 130; knowledge, 65, 78–102; labor scarcity, 115–19; market demand, 113–15; patents, 69–71, 113, 115, 119–124; play, 66–78; psychological factors, 64–5; rejection of, 130, 154; resource scarcity, 112–13; socioeconomic factors, 65, 103, 129
nuclear airplane, 183–4
nuclear energy, 161–8; Manhattan project, 162, 168; reactors, 161, 162, 163, 164, 165, 166, 167; Shippingport generating station, 164–7
nuclear merchant ship, 184–5
nuclear rocket, 182–3
nuclear submarine, 163–4, 166

Oersted, Hans Christian, 41
Ogburn, William F., 21–2, 23, 24, 114
O'Neale, Lila M., 104
Ortega y Gasset, José, 13, 208
Orwell, George, 55
Osage orange (bois d'arc), 51–4
Otto, Nikolaus A., 6, 40, 198, 203

Page, Charles G., 41
Paleolithic culture, 18
paper, 170
Papin, Denis, 92–7
Parke-Davis, 125
Pascal, Blaise, 92
patent system, 60–1, 69, 71, 120–4
patents, 64, 69, 71, 113, 114, 115; diversity, 2, 64, 119–20, 124; Great Britain, 120, 122; inventor, 60–1, 71, 121; Netherlands, 122–3; Switzerland, 122–3; U.S.A., 69, 74, 120–1; U.S.S.R., 123–4
Perkin, William H., 125
Perry, Commodore Matthew C., 189
Pershing, General John J., 160–1
phonograph, 139–40, 141–2
Pitt-Rivers, General Augustus Henry, 15, 16–21, 24, 108
plastics, 108
play and technology, 66–78
pneumatics, 92, 95, 96, 102
Popov, A.S., 60
pottery, 104–5, 106, 107, 108, 186
prehistory, 213–14, 215
printing, 169, 170, 173, 176, 192; typography, 170, 192–5; xylography (block printing), 170, 192–5, 203
progress: organic evolution, 217–18; technological, 20, 130–2, 210–18
Prometheus, 11

radio telegraphy, 99–101, 141, 216–17
railroad, 90–1, 153, 190, 195–7, 203; atmospheric, 177–81; India, 80–1
Ramelli, Agostino, 67–9
reaper, mechanical, 63, 151–4
reaping, hand, 63, 151, 153
recorder, magnetic-tape, 86, 140–1
revolution: scientific, 26–7; technological, 26–8, 61
Reynolds, Terry S., 146
Ricci, Matteo, 194
Rickover, Admiral Hyman G., 163–6
rifles, 17
Righi, Augusto, 100
Rivers, W. H. R., 186
Robinson, W. Heath, 76
Roosevelt, Franklin D., 162
Rose, Henry M., 53
Rosenberg, Nathan, 112, 144, 203–4

saw, Japanese hand, 190–1
Schiff, Eric, 122
Schmookler, Jacob, 69, 113–15, 119
Schuyler, Colonel John, 82
science fiction, 15, 57
scythe, 151, 153
selection, 135, 204–6, 210; artifactual and natural compared, 135–9; in Chinese culture, 169–176; economic constraints, 143–158; fads and fashions, 176–185; military necessity, 158–68; unconscious, 18
semiconductor, 44
ship, 22–3
Shockley, William, 44
Silberston, Z. A., 122
silk production, 83–6
skeuomorph, 106–8
sledge, 8, 212
Smiles, Samuel, 59
Sony, 86–7, 140–1
Sparks, Samuel S., 69
species: organic, 1, 138; technological, 2, 137–9
Spencer, Herbert, 17
spinning mule, 111
Stagenkunst, 149
Stanley steamers, 202
steamboat: American, 89–90; in India, 79–80
steam engine, 35–40, 41, 43, 141, 217; diffusion of, 82; Newcomen (atmospheric), 35, 37, 40, 92–7, 149–50; Watt, 35, 37, 40, 41, 96, 150
Stirling, Robert, 40
supersonic transport (SST), 154–8
Swan, Sir Joseph W., 60
Swift, Jonathan, 55–7
Szilard, Leo, 161–2

Taintner, Charles S., 141
Taylor, C. T., 122
Taylor, Theodore, 182
technology: autonomous, 204–6; Chinese, 169–76; military, 158–9; prehistoric, 109–10
technology transfer, 78–91; environmental influences, 88–91; imperialism, 79–81; India, 79–81; migration, 81–3; textile technology, 82–3
telegraphy, India, 80
television, 142, 217
textile machinery, 28, 82–3, 110–12
Theatrum machinarum, 67
Thorndike, Lynn, 130
Tikopia, 65–6
tools: hand, 190–1; metal, 31–2; stone, 13, 27, 30–2, 50, 104, 137
Torricelli, Evangelista, 92, 102
transistor, 43–6, 86–7, 217; Japan, 86–7; junction type, 45; point-contact type, 45; radio, 86
turbojet engine, 28–9
typewriter, 142

Usher, Abbott P., 21, 23–4
utility, 2

vacuum tube, 44, 45, 46
variations, random, 103–4
Vergil, Polydore, 129
Verne, Jules, 76
Villa, Pancho, 160
Villard d'Honnecourt, 73
visions, technological, 71–3

watch, digital, 59
waterwheel, 144–51; Antiquity, 145–6; Middle Ages, 147–9; post-Renaissance, 149, 150
Watt, James, 35, 37, 40, 63, 82, 150, 217
Wells, H. G., 76
Western Electric, 86
wheel, 7–11, 12
White Jr., Lynn, 133
Whitney, Eli, 32–4, 57, 61, 62, 63
windmill, 79
Winner, Langdon, 204–5
wireless telegraphy, 98–102, 217
wood-carvers, African, 105, 106
wool-combing, 111–12
World War I, 7, 160, 161
World War II, 161, 162, 168
Wright, Orville and Wilbur, 142, 217

Yurok-Karok Indians, 104

Zonca, Vittorio, 84